RESIDUE REVIEWS

VOLUME 54

RESIDUE REVIEWS

Residues of Pesticides and Other Contaminants in the Total Environment

Editor
FRANCIS A. GUNTHER

Assistant Editor
JANE DAVIES GUNTHER

Riverside, California

VOLUME 54

SPRINGER-VERLAG
NEW YORK • HEIDELBERG • BERLIN
1975

Library of Congress Catalog Card Number 62-18595.
Printed in the United States of America.

New York: 175 Fifth Avenue, New York, N.Y. 10010
Heidelberg: 6900 Heidelberg 1, Postfach 1780, West Germany

ISBN 0-387-90099-3 Springer-Verlag New York Heidelberg Berlin
ISBN 3-540-90099-3 Springer-Verlag Berlin Heidelberg New York

Foreword

Worldwide concern in scientific, industrial, and governmental communities over traces of toxic chemicals in foodstuffs and in both abiotic and biotic environments has justified the present triumvirate of specialized publications in this field: comprehensive reviews, rapidly published progress reports, and archival documentations. These three publications are integrated and scheduled to provide in international communication the coherency essential for nonduplicative and current progress in a field as dynamic and complex as environmental contamination and toxicology. Until now there has been no journal or other publication series reserved exclusively for the diversified literature on "toxic" chemicals in our foods, our feeds, our geographical surroundings, our domestic animals, our wildlife, and ourselves. Around the world immense efforts and many talents have been mobilized to technical and other evaluations of natures, locales, magnitudes, fates, and toxicology of the persisting residues of these chemicals loosed upon the world. Among the sequelae of this broad new emphasis has been an inescapable need for an articulated set of authoritative publications where one could expect to find the latest important world literature produced by this emerging area of science together with documentation of pertinent ancillary legislation.

The research director and the legislative or administrative advisor do not have the time even to scan the large number of technical publications that might contain articles important to current responsibility; these individuals need the background provided by detailed reviews plus an assured awareness of newly developing information, all with minimum time for literature searching. Similarly, the scientist assigned or attracted to a new problem has the requirements of gleaning all literature pertinent to his task, publishing quickly new developments or important new experimental details to inform others of findings that might alter their own efforts, and eventually publishing all his supporting data and conclusions for archival purposes.

The end result of this concern over these chores and responsibilities and with uniform, encompassing, and timely publication outlets in the field of environmental contamination and toxicology is the Springer-Verlag (Heidelberg and New York) triumvirate:

Residue Reviews (vol. 1 in 1962) for basically detailed review articles concerned with any aspects of residues of pesticides and other chemical contaminants in the total environment, including toxicological considerations and consequences.

Bulletin of Environmental Contamination and Toxicology (vol. 1 in 1966) for rapid publication of short reports of significant advances and discoveries in the fields of air, soil, water, and food contamination and pollution as well as methodology and other disciplines concerned with the introduction, presence, and effects of toxicants in the total environment.

Archives of Environmental Contamination and Toxicology (vol. 1 in 1973) for important complete articles emphasizing and describing original experimental or theoretical research work pertaining to the scientific aspects of chemical contaminants in the environment.

Manuscripts for *Residue Reviews* and the *Archives* are in identical formats and are subject to review, by workers in the field, for adequacy and value; manuscripts for the *Bulletin* are not reviewed and are published by photo-offset to provide the latest results without delay. The individual editors of these three publications comprise the Joint Coordinating Board of Editors with referral within the Board of manuscripts submitted to one publication but deemed by major emphasis or length more suitable for one of the others.

November 4, 1974 Joint Coordinating Board of Editors

Preface

That residues of pesticide and other contaminants in the total environment are of concern to everyone everywhere is attested by the reception accorded previous volumes of "Residue Reviews" and by the gratifying enthusiasm, sincerity, and efforts shown by all the individuals from whom manuscripts have been solicited. Despite much propaganda to the contrary, there can never be any serious question that pest-control chemicals and food-additive chemicals are essential to adequate food production, manufacture, marketing, and storage, yet without continuing surveillance and intelligent control some of those that persist in our foodstuffs could at times conceivably endanger the public health. Ensuring safety-in-use of these many chemicals is a dynamic challenge, for established ones are continually being displaced by newly developed ones more acceptable to food technologists, pharmacologists, toxicologists, and changing pest-control requirements in progressive food-producing economies.

These matters are of genuine concern to increasing numbers of governmental agencies and legislative bodies around the world, for some of these chemicals have resulted in a few mishaps from improper use. Adequate safety-in-use evaluations of any of these chemicals persisting into our foodstuffs are not simple matters, and they incorporate the considered judgments of many individuals highly trained in a variety of complex biological, chemical, food technological, medical, pharmacological, and toxicological disciplines.

It is hoped that "Residue Reviews" will continue to serve as an integrating factor both in focusing attention upon those many residue matters requiring further attention and in collating for variously trained readers present knowledge in specific important areas of residue and related endeavors involved with other chemical contaminants in the total environment. The contents of this and previous volumes of "Residue Reviews" illustrate these objectives. Since manuscripts are published in the order in which they are received in final form, it may seem that some important aspects of residue analytical chemistry, biochemistry, human and animal medicine, legislation, pharmacology, physiology, regulation, and toxicology are being neglected; to the contrary, these apparent omissions are recognized, and some pertinent manuscripts are in preparation. However, the field is so large and the interests in it are so varied that the editors and the Advisory Board earnestly solicit suggestions of topics and authors to help make this international book-series even more useful and informative.

"Residue Reviews" attempts to provide concise, critical reviews of timely advances, philosophy, and significant areas of accomplished or needed endeavor in the total field of residues of these and other foreign chemicals in any segment of the environment. These reviews are either general or specific, but properly they may lie in the domains of analytical chemistry and its methodology, biochemistry, human and animal medicine, legislation, pharmacology, physiology, regulation, and toxicology; certain affairs in the realm of food technology concerned specifically with pesticide and other food-additive problems are also appropriate subject matter. The justification for the preparation of any review for this book-series is that it deals with some aspect of the many real problems arising from the presence of any "foreign" chemicals in our surroundings. Thus, manuscripts may encompass those matters, in any country, which are involved in allowing pesticide and other plant-protecting chemicals to be used safely in producing, storing, and shipping crops. Added plant or animal pest-control chemicals or their metabolites that may persist into meat and other edible animal products (milk and milk products, eggs, etc.) are also residues and are within this scope. The so-called food additives (substances deliberately added to foods for flavor, odor, appearance, etc., as well as those inadvertently added during manufacture, packaging, distribution, storage, etc.) are also considered suitable review material. In addition, contaminant chemicals added in any manner to air, water, soil or plant or animal life are within this purview and these objectives.

Manuscripts are normally contributed by invitation but suggested topics are welcome. Preliminary communication with the editors is necessary before volunteered reviews are submitted in manuscript form.

Department of Entomology
University of California
Riverside, California
November 4, 1974

F.A.G.
J.D.G.

Table of Contents

Phthalate esters: Occurrence and biological effects

By

DAVID B. PEAKALL*

Contents

I. Introduction

This review considers the occurrence and biological activity of phthalates. The analytical procedures and problems have been reviewed by FISHBEIN and ALBRO (1972). Source documents of especial value were the National Institutes of Environmental Health Sciences Symposium (third issue of *Environmental Health Perspectives,* January 1973), Society of Plastic Engineers Symposium (March 1973), and the Toxicology Information Response Center Report (AUTIAN 1972 and 1973). Many articles in Russian and other eastern European

* Section of Ecology and Systematics, Langmuir Laboratory, Cornell University, Ithaca, New York 14850.

Table I. *U.S. production of individual phthalic esters, 1961–1970* (in thousands of pounds).

Phthalic ester	1970	1969	1968	1967	1966	1965	1964	1963	1962	1961
Butyl octyl	11,934	13,264	—	15,678	17,222	15,052	12,422	17,903	17,833	15,997
Dibutyl	22,919	34,476	29,466	22,801	20,236	20,012	18,223	18,183	15,486	15,178
Dicyclohexyl	4,537	4,412	5,189	—	6,739	7,719	6,730	6,543	6,369	5,557
Diethyl	20,636	22,460	22,722	20,830	21,451	17,999	15,053	14,602	16,142	17,332
Diisodecyl	123,359	136,708	136,793	123,133	103,131	89,552	77,963	66,914	62,680	48,345
Dimethyl	8,117	7,338	6,481	4,549	4,433	4,408	4,253	3,843	4,562	4,121
Dioctyl[a]										
Di(2-ethylhexyl)	350,396	355,345	330,484	293,243	253,000	212,360	188,761	175,398	171,037	138,330
Diiso-octyl	85,128	83,023	94,186	98,066	103,000	121,589	134,853	100,431	63,851	43,176
Di-tridecyl	15,130	22,057	17,735	18,209	19,379	12,888	11,523	7,643	4,398	2,747
n-Hexyl *n*-decyl	9,276	12,832	12,428	—	—	—	—	—	—	—
n-Octyl *n*-decyl	58,938	54,516	43,906	40,721	35,036	25,416	21,475	17,823	17,474	16,987
Other esters	144,690	137,401	141,238	138,414	170,586	150,684	110,127	92,660	90,163	68,741
Total	855,060	883,832	840,628	783,876	754,473	678,679	601,403	521,943	469,996	376,511

[a] The preliminary figure for dioctyl phthalates [di(2-ethylhexyl) and diiso-octyl] for 1971 was 431,884,000 lb (one percent less than 1970) and for 1972 was 450,801,000 lb (3.5 percent more than 1970).

languages have been obtained through the U.S.-U.S.S.R. Health Exchange Program under the sponsorship of the National Institutes of Environmental Health Sciences; these articles are considered in more detail as they are not readily available to many scientists

II. Production

a) United States

Detailed figures are available for the United States from the U.S. Tariff Commission Reports. These are given in Tables I and II. It will be seen that the most common ester is di(2-ethylhexyl)phthalate (DEHP) and that this ester has been widely used since 1949. In 1945 the most common ester was dibutyl and its major use was as a plasticizer for smokeless powder; dimethyl was used as an insect repellent. The rapid rise in the use of phthalates can be seen from Table II, although the figures for DEHP suggest that there has been little increase during the period 1969–1972. The total cumulative production of phthalic esters in the United States up to and including 1972 can be calculated as 12,500 million pounds.

b) Europe

The production figures of phthalic anhydride for the major industrial countries of western Europe are given in Table III. These figures

Table II. *Summary of U.S. production of phthalic esters* (in thousands of pounds).

Phthalic ester	1965–69	1960–64	1955–59	1950–54	1945–49
Butyl octyl	61,636[a]	77,345	[b]	[b]	[b]
Dibutyl	126,991	69,651	91,609	99,829	109,097[c]
Diethyl	89,262	79,901	83,303	79,713	37,017
Diisodecyl	466,383	291,440	97,517	[d]	[d]
Dimethyl	20,609	20,165	18,082	16,095	28,289[e]
Di(2-ethylhexyl)	1,444,432	821,340	409,144	235,607	[f]
Diiso-octyl	410,864	389,140	181,441	71,471	3,082[g]
Total, all esters	3,941,488	2,314,594	1,245,347	785,543	401,929

[a] Excludes 1968.
[b] Individual figures not given before 1960.
[c] Includes 45 million pounds in 1945 used mainly in smokeless powder.
[d] Individual figures not given before 1955.
[e] Includes 18.8 million pounds in 1945 used mainly in insecticide preparations.
[f] Individual figures not given before 1949.
[g] 1948 and 1949 only

Table III. *Production of phthalic anhydride in western Europe* (in millions of pounds).[a]

Country	1963	1964	1965	1966	1967	1968	1969
France	60	71	111	116	125	—	—
Germany	173	200	207	223	303	386	—
Great Britain	—	—	119	—	—	—	—
Italy	—	131	133	141	152	171	183

[a] Source, *Chemicals.*

are for the anhydride rather than the final esters as given for the U.S. production in Table I. The figures for West Germany give a ratio of 1:1.27 for phthalic anhydride to phthalic plasticizers. The total production of West Germany, France, Italy, and Great Britain for 1965 comes to 560 million pounds, which is roughly equivalent to the U.S. production of 608 million pounds. The rate of increase in Great Britain was stated to be ten percent per annum. The current production of phthalic esters in western Europe is probably somewhat in excess of that of the United States.

The total production of plastics and resins in the U.S.S.R. in 1965 was 1,600 million pounds (Hemy 1971). This is about 15 percent of U.S. production for the same year. It is not known what proportion of the plastics in the U.S.S.R. contained phthalates.

c) *Japan*

The 1970 Japanese production of polyvinyl chloride was 1,800 million pounds (*Japanese Plastics Industry Annual* 1970) or about 60 percent of U.S. production of this class of plastics.

d) *World*

It is likely that the world production of phthalate esters is three to four times the U.S. production, *viz.* 3,000 to 4,000 million pounds.

III. Manufacture and uses

Phthalate esters are manufactured from phthalic anhydride and the appropriate alcohol. Phthalic anhydride is manufactured from either naphthalene or *o*-xylene by oxidation and the alcohols by either the Oxo or the Ald-Ox process. The purity of the esters is stated to be 99.70 to 99.97 percent (Graham 1973). The main impurities are isophthalic acid, terephthalic acid, and maleic anhydride; these compounds are presumably converted to the corresponding ester. It

Table IV. *Uses of phthalic esters in th United States* (GRAHAM 1973).

A. As plasticizers

	(Millions of pounds)
Building and construction	
Wire and cable	185
Flooring	150
Swimming pool liners	20
Miscellaneous	32
Subtotal	387
Home furnishings	
Furniture upholstery	90
Wall coverings	38
Houseware	30
Miscellaneous	45
Subtotal	203
Cars (upholstery, tops, etc.)	114
Wearing apparel	72
Food wrapping and closures	25
Medical tubing and intravenous bags	21
Total as plasticizers	922
B. As nonplasticizers	
Pesticide carriers	—
Oils	—
Insect repellent	—
Total as nonplasticizers	50
Grand total	972

seems unlikely that traces of these materials could influence any biological effect that phthalates might exert.

The main uses of plastics containing phthalates are listed in Table IV. The possible environmental consequences of these uses are considered later.

IV. Physical properties

The vapor pressures and solubilities in water of some of the common phthalates are given in Table V. The relationship of these data to environmental movement of these compounds is considered later.

V. Natural occurrence

Phthalates have been reported in a wide variety of substances (oil, soil, plants, and animals) over a wide geographic area. This information is summarized in Table VI. Two questions arise: (1) Are the phthalates artifacts of the procedure used? (2) If not, are the phthalates truly native or are they or exogenous origin, absorbed by the plant of animal?

Undoubtedly, some of the results listed in Table VI are due to

Table V. *Some physical properties of phthalate esters* (PATTY 1967).

Phthalate ester	Boiling point (°C)	Vapor pressure (mm Hg)[a]			Solubility in water (g/100 ml)
		150°C	200°C	250°C	
Dicyclohexyl	220–228	<0.1	1.5	29.0	<0.010 (25°C)[a]
Dibutyl	340	0.8	14.0	100.0	0.001 (30°C)[a] 0.450 (25°C)
Diethyl	296	6.3	55.0	260.0	0.100 (32°C)[a]
Diisodecyl	256[b]	—	1.1	8.0	<0.005 (25°C)[a]
Di(2-ethylhexyl)	387[c]		1.2	18.0	0.005 (25°C)[a]
Dimethyl	282	13.0	91.0	360.0	0.400 (32°C)[a]

[a] Monsanto Chemical Co. data sheets.
[b] Ten mm Hg.
[c] Five mm Hg.

contamination during collection and analysis. Several of the authors refer to this as the likely source of the phthalates. BAUMAN *et al.* (1966) have shown that soil may be readily contaminated by collection in plastic bags. ETEMADI (1963) showed that butyl phthalate identified in a bacteria (*Corynebacterium parvum*) was an artifact derived from polyvinyl tubing. THIELE and TRUPER (1972) showed that phthalic esters were artifacts in bacterial lipids despite the fact that they avoided the use of plastics. These workers totally labeled chemolithotrophically grown bacteria with $^{14}CO_2$ and showed that the phthalic esters were not radioactive. If suitable precautions are taken and appropriate controls run, it is possible to eliminate the first possibility; however, it is often unclear whether or not suitable precautions were taken. The second question is much more difficult to answer. Some studies, *viz.* ASO *et al.* (1932) and SCHMID and KARRER (1945), were carried out before the widespread use of phthalates, making exogenous sources unlikely. The occurrence of phthalates in a wide variety of plants suggests natural occurrence. Detailed studies of plants from areas likely to be exposed to phthalate contamination and from remote areas would be valuable to obtain a baseline for the presumably natural occurrence of phthalates in plants. The occurrence in mammalian tissue has only been documented in the pineal gland and heart muscle of cattle; this may well be due to ingestion of plants containing phthalates.

VI. Toxicity

The low acute toxicity of phthalates to mammals is clearly shown by the data summarized in Table VII. The data on birds appear

Table VI. *Environmental occurrences of phthalates.*

Substance studied	Phthalate and concentration	Comments	Reference
Soil	Phthalic acid	—	Hayashi & Nagai (1960)
Soil (Prince Edward Is., Canada)	Phthalic acid (0.03% of total organic matter)	—	Schnitzer & Desjardine (1964)
Sediment (Ohio River, U.S.A.)	Phthalic diester	—	Cifrulak (1969)
Shale (Russia)	Phthalic acid	—	Kokurin & Galutkina (1967)
Crude oil (Italy, France)	Dioctylphthalate (0.15%)	"Reexamination with extreme care to prevent contamination proved that the ester is a true constituent of the oil"	Phillips & Breger (1958), Breger (1966)
Crude oil (U.S.A.)	Phthalic anhydride	—	Breger (1972)
Humis (Prince Edward Is., Canada)	Bis(2-ethylhexyl), dibutyl, dicyclohexyl, benzyl butyl (0.03% dry wt. of fulvic acid)	"Origin is uncertain"	Ogner & Schnitzer (1970), Matsuda & Schnitzer (1971)
Humis	Phthalic acid	—	Hayashi & Nagai (1960)
Azotobactor (*Azotobacter chroococcum*)	Phthalic acid	Produced in a culture	Aso *et al.* (1932)
Bacteria (*Mycobacterium tuberculosis*)	Phthalic acid 0.1% + ester (probably octyl)	—	Aebi *et al.* (1953)
Bacteria (*Corynebacterium diphtheriae*)	Dioctyl phthalate	Contamination unlikely	Asselineau (1961)
Fungus (*Gibberella fujikuroi*)	Phthalic acid, dimethyl phthalate ("Appreciable quantities")	—	Cross *et al.* (1963)
Fungus (*Alternaria kikuchiana*)	Diheptyl phthalate	—	Sugiyama *et al.* (1969)
Fungus (*Polyporus dryadeus*)	Phthalic acid, monomethyl phthalate	—	Batta & Rangaswami (1969)

Table VI. (*Continued*)

Substance studied	Phthalate and concentration	Comments	Reference
Plant (*Umbelliferae*) (*Ligusticum actilobum*)	Butylidene phthalide (0.0034%), butyl phthalide (0.0049%)	— — —	MITSUHASHI *et al.* (1963)
(*Apium graveolens*)	Butyl phthalide (0.0030%)	—	MITSUHASHI *et al.* (1963)
(*Levisticum officinale*)	Butylidene phthalide (0.0077%)	—	MITSUHASHI *et al.* (1963)
Plant (*Umbelliferae*)	Dimethyl, diisobutyl, iso-*n*-butyl, di-*n*-butyl, diisoamyl, di-*n*-amyl, di-*n*-hexyl, di-2-ethylhexyl	—	HAYASHI *et al.* (1967)
Plant (*Umbelliferae*)	Butylphthalide	—	BOHRMANN *et al.* (1967)
Cranberries (*Vaccinium vitis*)	Dimethyl (0.02%), diethyl (0.3%), diisobutyl (0.2%), dibutyl (2.8%)	All compounds may have originated in solvent	KLAUS & VON SYDOW (1967)
Raspberries	Phthalic acid	—	SCHINZ & SEIDEL (1957)
Poppies (*Papaver somniferum*)	Phthalic acid	Determination made before widespread use of these compounds	SCHMID & KARRER (1945)
Grapes (*Vitis vinifera*)	*n*-Butylphthalate (2.25 ppm of fresh wt.)	—	HAAGEN-SMIT *et al.* (1949)
	Dimethyl phthalate (identification not confirmed) + unidentified phthalate ester	—	STEVENS *et al.* (1966)
Tobacco	Phthalates (0.02% of dry leaf wt.)	Authors considered most materials identified were not artifacts	SWAIN *et al.* (1961)
	Phthalic acid (0.35 mg/100 g of tobacco smoked)	—	SCHMELTZ & SCHLOTZHAUER (1962)
	Phthalic acid (total of all acids, mainly succinic and phthalic, was 45–65 mg/100 g)	—	MOKHANACHEV & ASTAKHOVA (1968)

Oats	4-Hydrophthalide	—	Knight (1966)
Sake (rice beer)	Phthalic ester	—	Yamamoto & Omori (1964)
Corn oil	Di(2-ethylhexyl) phthalate	Considered formed by thermal oxidation, lack of optical activity favors exogenous origin	Beroza *et al.* (1968), Perkins (1967 a and b)
Olive oil	Phthalates (1–5 ppm)	—	Covello *et al.* (1967)
	Phthalates (1–2 ppm)	Higher in olive husk oil than olive oil, considered of natural origin	Ranaudo & Schettino (1969)
Kedwa oil (*Pandanus odoratissimus*)	Phthalic ester	—	Dhingra *et al.* (1954)
Eucalyptus oil (*Eucalyptus punctata*)	Phthalate (mp 200–201°C)	—	Proenca da Cunha (1966)
Lily of the valley oil (*Convallaria keisukei*)	Phthalic acid (small part of 0.13% methanol-soluble portion)	—	Wakayama & Namba (1965)
Fish meal lipids	Dioctyl phthalate	May be artifact	Sims (1967)
Butterfat	Dibutyl phthalate	Probably contamination from tubing	Parodi & Dunstan (1968)
Jellyfish (*Atolla* spp.) (N. Atlantic)	Phthalic acid (0.01% wet wt., 13% of lipids)	Considered present before capture but most unlikely a normal metabolite	Morris (1970)
Cockroach (*Periplaneta americana*)	Dibutylphthalate (1.75 ppm)	Reared under conditions that definitely exclude contamination	Jones *et al.* (1973)
Armyworm moth (*Spodoptera eridania*)	Di(2-ethylhexyl)	—	Jones (unpublished, cited in Jones *et al.* 1973)
Cattle pineal gland	Dioctyl phthalate[b]	Not due to contamination after collection	Taborsky (1967)
Bovine heart muscle	Diethylhexyl phthalate	Not due to contamination after collection	Nazir *et al.* (1967), Nazir *et al.* (1971)

[a] Percent of essential oil which is 1.1 ppm of fresh weight.
[b] Considered by Beroza *et al.* (1968) to be diethylhexyl phthalate.

Table VII. *Acute toxicities of phthalates.*[a]

Phthalate	Species	Route	LD_{50}	Reference
Dimethyl	Mouse	oral	7.2 ml/kg	DRAIZE ET AL. (1948)
	Mouse	i.p.	1.6 g/kg	CALLEY *et al.* (1966)
	Mouse	i.p.	2.4 ml/kg	HODGE *et al.* (1942)
	Mouse	i.p.	3.6 g/kg	KAREL *et al.* (1947)
	Rat	oral	6.9 ml/kg	DRAIZE *et al.* (1948)
	Rat	i.p.	3.4 ml/kg	SINGH *et al.* (1972)
	Rabbit	oral	4.4 ml/kg	DRAIZE *et al.* (1948)
	Rabbit	dermal	>10 ml/kg	
	Guinea pig	oral	2.4 ml/kg	
	Chicken	oral	8.5 ml/kg	
Diethyl	Mouse	i.p.	2.8 g/kg	CALLEY *et al.* (1966)
	Mouse	i.p.	2.75 g/kg	KAREL *et al.* (1947)
	Rat	i.p.	5.1 ml/kg	SINGH *et al.* (1972)
	Rat	oral	8.2 ml/kg	TYSON (1972)
	Rabbit	oral	1.0 g/kg	KEMP (1934)
	Rabbit	i.v.	0.1 g/kg	
Dibutyl	Mouse	oral	<13.0 g/kg	BOURNE *et al.* (1968)
	Mouse	i.p.	4.0 g/kg	CALLEY *et al.* (1966)
	Mouse	i.p.	21.5 g/kg	MASLENKO (1968)
	Mouse	i.p.	7 ml/kg	ELIZAROVA (1961)
	Mouse	i.p.	5.5 mg/kg	HODGE *et al.* (1942)
	Mouse	i.p.	4.1 g/kg	KAREL *et al.* (1947)
	Rat	oral	23 g/kg	RADEVA & DINOEVA (1966)
	Rat	oral	17.9 g/kg	MASLENKO (1968)
	Rat	oral	13.8 ml/kg	HOMROWSKI & NIKONOROW (1959)
	Rat	oral	>20 g/kg	LEHMAN (1955)
	Rat	oral	8 g/kg	SMITH (1953)
	Rat	oral	8 g/kg	BOURNE *et al.* (1968)
	Rat	inh.	0.7–0.9 mg/l.	SPASOVSKI (1964)
	Rat	s.c.	<6.2 g/kg	KOWALSKI & BASSENDOWSKA (1965)
	Rat	i.p.	4.5 g/kg	
	Rat	i.p.	10 ml/kg	ELIZAROVA (1961)
	Rat	i.p.	3.05 ml/kg	SINGH *et al.* (1972)
	Rat	i.p.	8.3 ml/kg	HOMROWSKI & NIKONOROW (1959)
	Bluegill	LC_{50}[b]	1.23 mg/l.	MAYER & SANDERS (1973)
	Channel catfish	LC_{50}[b]	3.72 mg/l.	
	Scud	LC_{50}[b]	7.00 mg/l.	
Diisobutyl	Mouse	oral	>12.8 g/kg	FASSETT (1967)
	Mouse	i.p.	4.5 g/kg	CALLEY *et al.* (1966)
	Rat	oral	20–25 g/kg	FASSETT (1967)
	Rat	i.p.	3.75 ml/kg	SINGH *et al.* (1972)
	Guinea pig	dermal	10 ml/kg	FASSETT (1967)
Dihexyl	Rat (male)	oral	29.6 ml/kg	KRAUSKOPF (1973)
	Rat (female)	oral	38.9 ml/kg	SMYTH *et al.* (1954)
	Rat	oral	29.6 g/kg	
	Rabbit	dermal	20 ml/kg	

Table VII. (*Continued*)

Phthalate	Species	Route	LD_{50}	Reference
Di(2-ethyl-hexyl)	Mouse	oral	>128 g/kg	HODGE (1943)
	Mouse	i.p.	14.2 g/kg	CALLEY *et al.* (1966)
	Mouse	i.p.	>75 g/kg	GESLER & KARTINOS (1970)
	Rat	oral	>26 g/kg	FASSETT (1967)
	Rat	oral	30.6 g/kg	SHAFFER *et al.* (1945)
	Rat	oral	>34 g/kg	HODGE (1943)
	Rat	i.p.	>50 ml/kg	SINGH *et al.* (1972)
	Rabbit	oral	33.9 g/kg	SHAFFER *et al.* (1945)
	Guinea pig	dermal	>10 g/kg	FASSETT (1967)
	Mouse	oral	>13 g/kg	
Dioctyl	Rat	oral	45.1 ml/kg	HOMROWSKI & NIKONOROW (1959)
	Rat	oral	30 g/kg	RADEVA & DINOEVA (1966)
	Rat	i.p.	46 ml/kg	HODGE *et al.* (1942)
	Rat	i.p.	50 ml/kg	HOMROWSKI & NIKONOROW (1959)
	Rat	i.p.	>50 ml/kg	SINGH *et al.* (1972)
	Guinea pig	dermal	>5 ml/kg	FASSETT (1967)
Dicapryl	Mouse	i.p.	14.2 g/kg	CALLEY *et al.* (1972)
Diisonoyl	Rat	oral	>10 g/kg	LIVINGSTON (1971)
Diisodecyl	Rat	oral	>64 ml/kg	SMYTH *et al.* (1954)
Ditridecyl	Rat	oral	>64 ml/kg	
Butylbenzyl	Mouse	i.p.	3.2 g/kg	CALLEY *et al.* (1972)
	Rat	i.p.	1.8 g/kg	MALLETTE (1952)
	Rat	oral	4.0 g/kg	
Dialkyl 79 phthalate	Mouse	oral	>20 g/kg	GAUNT *et al.* (1968)
	Mouse	i.p.	>20 g/kg	
	Rat	oral	>20 g/kg	
	Rat	i.p.	>20 g/kg	
	Rat	i.p.	>44 g/kg	ELIZAROVA (1961)
	Rat	oral	13.3 ml/kg	BROWN *et al.* (1970)
	Mouse	oral	7 ml/kg	
Dialkyl 9–11 phthalate	Rat	oral	>20 ml/kg	
	Mouse	oral	>20 ml/kg	
Dimethoxy-ethyl	Mouse	oral	3.2–6.4 g/kg	FASSETT (1967)
	Mouse	i.p.	2.51 g/kg	CALLEY *et al.* (1966)
	Rat	oral	>4.4 g/kg	FASSETT (1967)
	Rat	i.p,	3.7 ml/kg	SINGH *et al.* (1972)
	Guinea pig	oral	1.6–3.2 g/kg	FASSETT (1967)
	Guinea pig	dermal	>10 ml/kg	FASSETT (1967)
Diallyl	Mouse	i.p.	0.7 g/kg	MCOMIE (1949)
	Mouse	i.p.	0.6 ml/kg	MCOMIE & ANDERSON (1947)
	Rat	oral	0.8–1.7 g/kg	MCOMIE (1949)
	Rabbit	oral	1.7 g/kg	MCOMIE (1946)
	Rabbit	dermal	3.4 ml/kg	MCOMIE (1949)
Butyl carbo-butomethyl	Rat	oral	14.6 ml/kg	PATTY (1967)
	Rat	i.p.	6.89 g/kg	SINGH *et al.* (1972)

[a] Some additional information on mixed esters is given in KRAUSKOPF (1973).
[b] Twenty-four hours.

Table VIII. *Acute toxicity of dibutyl and dioctyl phthalates to rats* (HOMROWSKI and NIKONOROW 1959).

Phthalate	Observation period	Method of administration	LD_{50} (ml/kg)
Dibutyl	24 hours	oral	13.8
	24 hours	peritoneal	8.3
	7 days	oral	11.9
	7 days	peritoneal	6.8
Dioctyl	24 hours	oral	45.1
	24 hours	peritoneal	50.5
	7 days	oral	33.7
	7 days	peritoneal	37.3

to be limited to one species and one phthalate (DRAIZE *et al.* 1948), and in this case the acute toxicity was of the same order as that found for mammals. The values obtained for fish are much higher (*i.e.*, less toxic) than those found for the common organochlorine pesticides.

Toxicity studies go back 50 years to SMITH (1924), who found that rats were able to survive 0.5 percent of their body weight of diethyl phthalate and pullets as much as 1.5 percent. BLICKENSDORFER and TEMPLETON (1930) found that diethyl phthalate at doses of two to three ml/kg/day could be tolerated by rabbits due to rapid excretion of this material. Much of the more recent data on chronic ingestion studies have been ably summarized by KRAUSKOPF (1973) and it is not proposed to resummarize these findings[1]. It is proposed to detail some of the work from the eastern European literature and relate it to the material presented by KRAUSKOPF. Some of the more subtle effects noted are considered under the heading of physiological effects.

PIEKACZ (1971 a) reviewed the literature on the toxicity of dibutyl phthalate and dioctyl phthalate as a preliminary to more detailed studies. The starting point for these studies was the findings of HOMROWSKI and NIKONOROW (1959), given in Table VIII. PIEKACZ (1971 b) studied the effect of one, five, and ten percent of the LD_{50} values given daily by stomach tube for three months and of one percent given in the diet for seven and 12 months for both dibutyl and dioctyl phthalates on rats. Gross effects were seen only at ten percent LD_{50} (3.50 g/kg) dioctyl phthalate; these rats showed uncertainty of movement, sluggishness, diarrhea, and death (15 out of

[1] One additional long-term study on Alkyl phenyl phthalate should be mentioned. BREWER *et al.* (1955) found that dietary levels up to four percent did not affect growth or life-span of rats over a two-year period, nor were dogs affected over a one-year period.

20 died within three months). A statistically significant lower body weight was found for rats on the one percent LD_{50} diet of dioctyl phthalate at seven months, but this decrease was not found at 12 months. Increased liver weights were found in the experiments using dioctyl phthalate and most of these increases were statistically significant (see Table IX). The figures for dibutyl phthalate suggest an initial increase of liver size followed by a fall to or below control values. No histopathological differences were established for either compound. CARPENTER *et al.* (1953) and HARRIS *et al.* (1956) also noted changes in liver weights in the same dose range, although the first-mentioned workers found this only with females. No significant changes were noted in the weights of the kidney or spleen with either compound.

A statistically significant increase in the activity of serum alanine aminotransferase was noted after exposure to one and five percent LD_{50} dosage of dioctyl phthalate for three months. Insignificant increases were noted with the same dosage of dibutyl phthalate. No significant increases were found by either dibutyl or dioctyl phthalate on the serum levels of asparagine aminotransferase at dosages of one or five percent LD_{50} for three months. However, dioctyl phthalate (one percent LD_{50} for one year) caused increases of both analine and asparagine aminotransferases, whereas no effect was caused by dibutyl phthalate under these conditions. No variation from normal values was observed in the red or white cell count.

PIEKACZ (1971 c) examined the effect of one percent and five percent dioctyl and dibutyl phthalate on reproduction and fetal development in the rat. Two schemes of administration were used:

Table IX. *Effects of dibutyl and dioctyl phthalates on the liver weights of rats* (PIEKACZ 1971 b).

Phthalate	Dose (% LD_{50}) and duration	Liver weight as % of body weight	
		Male	Female
Control	0	3.08 ± 0.77	3.23 ± 0.77
Dioctyl	1%, 3 mo.	3.71 ± 0.70[a]	3.57 ± 0.84[a]
Dibutyl	1%, 3 mo.	3.49 ± 0.84[a]	3.48 ± 1.15
	10%, 3 mo.	3.88 ± 0.64[a]	4.49 ± 1.06[a]
Control	0	3.45 ± 1.14	3.41 ± 1.58
Dioctyl	1%, 7 mo.	3.94 ± 1.16	4.07 ± 0.69[a]
Dibutyl	1%, 7 mo.	3.21 ± 0.55	2.97 ± 1.16
Control	0	3.11 ± 1.11	3.09 ± 0.64
Dioctyl	1%, 12 mo.	3.57 ± 0.97	4.53 ± 1.44[a]
Dibutyl	1%, 12 mo.	2.88 ± 0.80	3.20 ± 0.72

[a] Significant at 95 percent level.

Table X. *Effect of dioctyl and dibutyl phthalates (for three months before conception) on reproduction in the rat* (PIEKACZ 1971 c).

Phthalate	Fetuses			Av. wt. of placenta (g)
	No. produced	No. reabsorbed	Av. wt. (g)	
Control	98	5	3.38 ± 0.29	0.43 ± 0.07
Dioctyl				
1% LD_{50}	77	6	3.50 ± 1.31	0.47 ± 0.16[a]
5% LD_{50}	101	2	3.35 ± 0.50	0.53 ± 0.13[a]
Dibutyl				
1% LD_{50}	94	8	3.59 ± 0.54	0.45 ± 0.06
5% LD_{50}	68[a]	13	3.68 ± 1.03	0.52 + 0.16

[a] Difference significant at five percent level.

(1) daily for three months before conception and (2) daily for 21 days after conception. These results are summarized in Tables X and XI.

DUEVA and ALDYREVA (1966) studied the sensitizing and irritating action of dibutyl, butyl benzyl, dioctyl, and higher esters (chain length not listed) on guinea pigs. These workers applied 100 mg of the phthalate to two sq cm of sheared skin daily for 20 days. The effects were most pronounced with the higher esters. Histological studies did not show any change to the internal organs, but irritation of the reticuloendothelial cells of the skin was noted.

RALL (1973) considers that most of man's exposure to phthalates is by inhalation rather than the oral route, a viewpoint supported

Table XI. *Effect of dioctyl and dibutyl phthalates (for 21 days after conception) on reproduction in the rat* (PIEKACZ 1971 c).

Phthalate	Fetuses			Av. wt. of placenta (g)
	No. produced	No. reabosrbed	Av. wt. (g)	
Control	91	4	3.76 ± 0.70	0.52 ± 0.07
Dioctyl				
1% LD_{50}	99	11	3.38 ± 0.47[a]	0.49 ± 0.06
5% LD_{50}	98	13	3.07 ± 0.61[a]	0.48 ± 0.07[a]
Dibutyl				
1% LD_{50}	106	2	3.56 ± 1.08	0.44 ± 0.07[a]
5% LD_{50}	81	22	2.92 ± 0.47[a]	0.47 ± 0.11[a]

[a] Difference significant at five percent level

by calculations of probable environment inputs (Table XIV). The data on the effect of inhalation appear to be limited to papers by SPASOVSKI (1964), DVOSKIN *et al.* (1969), and MEN'SHIKOVA (1971), all on dibutyl phthalate. In short-term experiments, SPASOVSKI found that only one mouse out of 22 died at concentrations of 0.43 to 0.61 mg/l., nine out of 18 at 0.70 to 0.93 mg/l., and that a concentration of 1.9 mg/l. was fatal to all mice. Chronic effects were studied by subjecting mice to a six hour/day exposure for 93 days. The concentrations to which the mice were subjected were varied during this time, the regime being 11 days at 0.078 to 0.085 mg/l., 15 days at 0.040 to 0.061 mg/l., 60 days at 0.017 to 0.040 mg/l., and a final six days at a higher concentration, *viz.*, 0.17 to 0.42 mg/l. No differences were noted in mortality or weight change (total, liver, and kidney). A small decrease in positive response of reflexes and a small increase in swimming ability were noted; neither was statistically significant. SPASOVSKI considered that the threshold concentration was 0.30 mg/l. and proposed a permissible standard for concentration (PSC) of 0.001 mg/l. (one mg/cu. m).

DVOSKIN *et al.* (1969) exposed rats for 2.5 months in chambers containing "pavinol," a plastic containing dibutyl phthalate, and made detailed studies of the blood chemistry (see Table XII). Six groups of 12 rats each were used: two control groups and two for each of two samples of plastic. In one case, the air was changed 15 times and once in the other. No dibutyl phthalate was detected with the 15 air changes and these results are not considered in Table XII. Statistical evaluation of these results is not given, but the authors comment on the decrease of the percent of albumin, the albumin-globulin ratio, the phagocytic index, and the increase of gamma globulin. Some behavior studies, positive and differentiated conditioned reflexes, were made. Considerable differences before and after in both controls, and experimental groups make evaluation difficult. In another experiment, rats were exposed to 0.2 and 0.4 mg/cu.m for 2.5 months. during two periods of this experiment, the experimental animals exhibited a transient weight loss which was significantly more marked than the controls. An increase of gamma globulin was noted at the higher concentration of dibutyl phthalate during the fourth and sixth weeks of the experiment. This group also showed alterations in the phagocytic activity of neutrophils after one month, but these values also returned to control levels.

MEN'SHIKOVA (1971) studied the effect of inhalation of dibutyl phthalate on rats by continuous exposure for 93 days. The concentrations studied were 0.098, 0.256, and 0.98 mg/cu.m, which are two to three orders of magnitude lower than those used by SPASOVSKI (1964), and no significant changes in behavior or weight were noted. Hemoglobin levels and red blood cell counts remained constant. MEN'SHIKOVA stated the white blood cell count deviated from control

Table XII. *Effects of inhalated dibutyl phthalate on the blood chemistry of rats* (after DVOSKIN *et al.* 1969).

Sample	Albumin		Gamma globulin		Albumin-globulin ratio		Phagocytic index	
	Before (%)	After (%)	Before (%)	After (%)	Before	After	Before	After
Control	44.5 ± 1.6	40.3 ± 2.9	13.1 ± 1.1	16.7 ± 1.8	0.6 ± 0.2	0.7 ± 0.1	1.9 ± 0.2	2.3 ± 0.3
Control	42.1 ± 2.2	40.8 ± 4.8	13.5 ± 1.6	14.2 ± 1.7	0.9 ± 0.1	0 7	2.2 ± 0.4	2.0 ± 0.1
"Pavinol" 2[a]	47.0 ± 2.6	41.0 ± 1.9	15.1 ± 2.1	13.6 ± 3.3	0.9 ± 0.1	0.6 ± 0.1	2.3 ± 0.2	2.1 ± 0.2
"Pavinol" 1[b]	54.2 ± 2.5	42.5 ± 4.1	9.8 ± 1.1	16.7 ± 2.5	1.2 ± 0.1	0.8 ± 0.1	3.1 ± 0.9	1.7 ± 0.2

[a] 1.2 mg/cu. m.
[b] 8 mg/cu. m.

values in the two groups with the highest concentrations, but his diagram of these results suggests that fluctuations of all groups, including controls, was too great for a clear-cut conclusion to be drawn. The increase of gamma globulin is more convincing and appears to be dose-dependent. These experiments were done in connection with assessing the possible effect of liberated phthalates in the living quarters of boats. MEN'SHIKOVA found that even three years after construction the levels ranged as high as 1.22 mg/cu.m. The olfactory threshold for human beings was found to range from 0.26 to 1.47 mg/cu.m. The maximum ineffective concentration on ocular sensitivity was found to be 0.11 mg/cu.m. The effect of dibutyl phthalate on the electrical activity of the cerebral cortex was investigated by the inculcation of an electrocortical conditioned reflex in three subjects. Concentrations of 0.12 and 0.15 mg/cu.m produced the electrocortical conditioned reflex in all three subjects, whereas a concentration of 0.093 mg/cu.m had no effect. MEN'SHIKOVA recommended a PSC of 0.1 mg/cu.m, or $\frac{1}{10}$ of that recommended earlier by SPASOVSKI.

The values reported by LORANSKII *et al.* (1970) for new rooms with PVC tiles are between proposed values, *viz.*, 0.15 to 0.26 mg/cu.m. On the other hand, the values reported by MILKOV *et al.* (1969) of ten to 66 mg/cu.m in the air of the work zone of prime-coat workers and 1.7 to 40 mg/cu.m for the mixes part of the production are greatly in excess of SPASOVSKI's PSC. These workers found a moderately pronounced increase in toxic polyneuritis, the frequency and degree of which increased with the length of stay on the job. Investigations of the sensory functions showed an early lowering of the excitability of the vestibular and olfactory responses.

In the United States, the Threshold Limit Value (TLV) has been set by the American Conference of Industrial Hygienists (1972) for dimethyl, dibutyl, and di(2-ethylhexyl) phthalates at five mg/cu.m. LEVINSKAS (1973) tabulates the U.S. data on inhalation. The longest period of exposure is eight hours, which leads LEVINSKAS to comment that "Experimental evidence for the long-term safety of vapor exposures to these compounds is scanty." Accepting SPASOVSKI's threshold value of 0.30 mg/l. (300 mg/cu.m) as the best available figure, the safety margin for DBP is 60. The work of DVOSKIN and co-workers and MEN'SHIKOVA on gamma globulin suggests that the no-effect level (which may be quite different from a safe level) may be around the United States' TLV.

MASLENKO (1967, 1968, and 1969) suggests a maximum permissible concentration of two mg/l. for dibutyl and 2.5 mg/l. for dioctyl phthalate in water. These figures are based on the olfactory threshold and effects on oxygen levels. He concludes that these values are below the effective dose for warm-blooded animals. While this is correct, the work of MAYER and SANDERS (1973) would predict massive fish kills at these concentrations.

VII. Physiological effects

a) *Carcinogenic effects*

Chronic toxicity studies, summarized by Krauskopf (1973) and this review, give no indication that any phthalates are carcinogens. Oppenheimer *et al.* (1953) reported on malignant tumors (the great majority fibrosarcomas) caused by commercial PVC. This study was expanded (Oppenheimer *et al.* 1955) to evaluate the effects of many polymers as carcinogenic agents in rats. Small pieces of film were inserted subcutaneously into the abdominal wall. Two films of interest here were polyvinyl chloride A (a commercial product known to contain additives) and polyvinyl chloride D (which contained no plasticizers, catalysts, or other additions). PVC A (plain film) caused malignant tumors in 38.6 percent of the rats, but none when a perforated film was used. This striking difference was not noted with other films. PVC D was only reported as in progress in the 1955 paper, with four malignant tumors produced in 20 months. Oppenheimer *et al.* (1955 and 1959) concluded, based on many chemically different films, that the primary carcinogen is the macromolecule itself rather than any additive or impurity which may be present in the commercial film. Unfortunately, this point is not rigorously proved in the case of PVC and their additives in this paper.

b) *Teratogenic effects*

Since the well-known teratogenic agent, thalidomide, has a phthalate moity, it is not surprising that possible teratogenic effects have been assayed by a number of workers. The vast majority of the work on thalidomide shows that the metabolites are not themselves teratogenic (Smith *et al.* 1965, Williams 1968, Verrett *et al.* 1969, Kohler *et al.* 1971). The reverse was found by Baroche (1965 and 1968) who reported a dose-dependent increase in malformations of *Drosophila* with potassium phthalate and phthalic acid.

Injection into the developing chicken egg has been used by several investigators to study the teratogenic effects of phthalates. The dosage is usually expressed in terms of the volume injected. In this connection, it can be calculated that 0.05 ml is roughly equivalent to 1,000 ppm, assuming uniform mixing within the egg. While this figure is rough, it can be used for comparison with environmental levels. Guess *et al.* (1967) reported the occurrence of deformities of the legs and adverse neurological effects when 0.1 ml of butyl octyl or butyl decyl phthalate was injected into chicken eggs. Haberman *et al.* (1968) found adverse neurological effects when 0.1 ml of butyl octyl (five cases out of 35), dioctyl (five cases out of 32), and octyl isodecyl (two cases out of 13) were injected. Embryonic abnormalities were noted for dioctyl (one case) and octyl isodecyl (two cases). No effects were noted with dibutyl, butyl benzyl, or diethyl phthalates. Complete failure to hatch was noted with dimethyl phthalate. BOWER *et al.* (1970) noted marked adverse effects with dibutyloxyethyl phthalate (four abnormalities out of 41 at 0.025 ml and ten out of 43 at 0.050 ml). Some effects, although less marked, were noted with di-2-methoxyethyl and *n*-octyl-*n*-decyl phthalates. The abnormalities noted were neuromuscular effects, *viz.*, tremors, swaying, and impairment in the ability to stand or walk. Congenital malformations were noted with dibutyloxyethyl phthalate, and in a single case with diethyl phthalate. No effects were seen with dibutyl phthalate. Increased mortality was noted with several phthalates (notably octyl isodecyl, di-2-methoxyethyl, and dibutyl phthalate) but the mortality figures of both inoculated and uninoculated controls were too high to give confidence in this phase of the experiments. MCLAUGHLIN *et al.* (1963) listed di(2-ethylhexyl) phthalate as a chemical with an intermediate order of toxicity on the basis of injection into chicken eggs. However, their results showed that the percent hatch was identical to controls, *viz.*, 95 percent at the one dose tested (0.05 ml).

The teratogenicity of phthalate esters in rats has been examined by SINGH *et al.* (1972). They administered the phthalates by intraperitoneal injection on the fifth, tenth, and fifteenth days of gestation. Six phthalates (*viz.*, dimethyl, dimethoxyethyl, diethyl, dibutyl, diisobutyl, and butyl carbobutoxymethyl) were administered at $\frac{1}{10}$, $\frac{1}{5}$, and $\frac{1}{3}$ of their acute LD_{50}'s; di(2-ethylhexyl) and dioctyl were given at five and ten ml/kg. The percentage of reabsorptions was increased in most, and fetus weight decreased in all cases. Gross abnormalities were most marked with dimethoxyethyl, and definitely increased with dioctyl and dimethyl. The number of skeletal abnormalities was increased in all cases except dioctyl and di(2-ethylhexyl), the effects being most marked with the more soluble compounds.

It is clear that phthalates are teratogenic at high dosages. Their importance or otherwise at current environmental levels remains to be assessed.

c) *Mutagenicity*

Unpublished work of SINGH *et al.* (cited in DILLINGHAM and AUTIAN 1973) shows some mutagenic effects of di(2-ethylhexyl) and dimethoxyethyl phthalates. These workers used the dominant lethal assay test in mice and found a decrease of implants and number of live fetuses for both phthalates at the highest doses tested, *viz.*, a single injection of 25.6 ml/kg of DEHP and 2.4 ml/kg of DMEP. Increased fetal deaths were noted at all concentrations of DEHP (12.6 to 25.6 ml/kg) and at the highest concentration of DMEP. As with teratogenic effects, the importance of these effects remains to be assessed.

d) *Other physiological effects*

GUESS *et al.* (1967) and HABERMAN *et al.* (1967 and 1968) studied the effect of plasticizers on the complement activity of guinea pig serum. They found that dimethyl and dioctyl phthalates decreased the complement activity but that dibutyl phthalate did not cause an effect. Similar effects were found with some heat stabilizers and inhibitors used in the manufacture of plastics. Other effects such as hemolysis of blood and clumping of cells were considered. All effects observed were at high concentrations.

NEMATOLLAHI *et al.* (1967) evaluated the effect of phthalates on mammalian L-cell cultures, chicken embryo cultures, and intraperitoneal (i.p.) injection into mice. In the first two experiments, 20 mg of solid or 35 mg of liquid phthalates were applied directly to the cultures. Calculation of effective concentration is difficult for these experiments. In the intact animal experiments, five mmol./kg was used. Good aggreement between the three methods was found, the effects decreasing as chain length increased.

CALLEY *et al.* (1966) found that i.p. doses of 0.5 g/kg had effects on hexobarbital narcosis. The sleeping time was increased for dimethyl, diethyl, dibutyl, diisobutyl, dimethoxyethyl, and butyl benzyl phthalates but decreased for di(2-ethylhexyl) and dicapryl phthalates. It was considered that some phthalates act as central nervous system depressors while others act as stimulators. PFUDERER and FRANCIS (1973) gave a preliminary report of depression of the heart rate of goldfish by an extract from Tennessee Valley Authority lakes. The compounds were identified as benzyl butyl and dioctyl phthalates. These findings were confirmed using known compounds. FRANK *et al.* (1963) found that phthalic acid at one mg/ml reduced the growth of protozoa (*Ochromonas danica*) twelve-fold.

PROTSENKO (1970) examined the effects of inhalation of phthalic anhydride on the gonads of rats. The main findings are detailed in Table XIII. The reduction in the concentration of absorbic and dehydroascorbic acids is marked and is dose-dependent. The possible physiological or biochemical consequences are not discussed. Little or no effect is seen on the nucleic acids. Motility time of spermatozoa for rats exposed to 1.0 mg of phthalic anhydride/cu.m was 40 minutes, only ½ the control value of 82 minutes. The time was reduced to 60 minutes for those rats exposed to 0.2 mg/cu.m and no effect was observed at 0.02 mg/cu.m. The statistical significance of these results was not discussed.

VIII. Plastics, blood storage, and other medical uses

The possible effects of extracting materials from plastics used in the storage and transmission of blood have interested many workers.

Table XIII. *Effects on rat gonads of inhalation exposure to phthalic anhydride for 45 days* (PROTSENKO 1970).

Sample (mg/cu. m)	Concentration (mg-%)			
	Absorbic acid	Dehydro-ascorbic acid	DNA	RNA
Control	31.25 ± 1.36	2.39 ± 0.11	3.24 ± 0.37	1.50 ± 0.32
0.02	29.51 ± 0.86	1.90 ± 0.13[a]	2.32 ± 0.30	1.78 ± 0.24
0.20	24.70 ± 0.57[b]	1.40 ± 0.14[b]	3.45 ± 0.20	1.72 ± 0.25
1.00	19.19 ± 0.48[c]	0.29 ± 0.05[c]	3.53 ± 0.20	2.43 ± 0.25[a]

[a] Significant at 95 percent level.
[b] Significant at 99 percent level.
[c] Significant at 99.9 percent level.

Concern appears to have surfaced independently in different countries at different times since plastic bags were first used to store blood in 1950 (WALTER 1951). The importance of having this information readily available is shown by the statement of TRIMBLE *et al.* (1966): "The energy and resources of a research team studying the hormonal theory of hibernation were diverted for over a year by chemical contamination from plastic tubing." The Pharmacological Institute in Stockholm considered the use of plastics in blood transfusion equipment (GULLBRING 1964) and the necessary chemical and biological tests to evaluate plastics (GULLBRING *et al.* 1964) nearly a decade ago. Similar methods of evaluating plastics for medical uses were proposed independently by GUESS and HABERMAN (1968). Concern in the United States was not stimulated until the work of JAEGER and RUBIN (1970 a).

MEIJLER[2] and DURRER (1959) noted premature cardiac arrest and ventricular fibrillation in the isolated perfused rat heart when blood was circulated through some types of PVC tubing. MEYLER[2] *et al.* (1960) stated that experiments were carried out with five types of PVC, but details are given for only two: one with 43.7 percent phthalate, probably di(2-ethylhexyl), with an organic tin stabilizer, and one with 34.5 percent phthalate with a barium-calcium stabilizer. Cardiotoxic effects were noted only with the PVC containing organic tin.

STEWART and STURRIDGE (1959) studied the effect of several different types of tubing on the degree of hemolysis of blood circulating through them. They found considerable hemolysis both with rubber

[2] Undoubtedly the same person, but the name is spelled differently in two journals.

and with PVC. No evidence was presented as to the causative agent. KEITH *et al.* (1961) reported on massive hemolysis due to a single segment of plastic tubing in an extracorporeal pump-oxygenator used in a cardiopulmonary bypass. Previously the same plastic (type not stated) had been used without incident. BRAUN and KUMMELL (1963) found that aqueous extracts of PVC caused hemolysis of red blood cells *in vitro*. No evidence was presented as to whether phthalates or some other material was the active agent. HIROSE *et al.* (1963) noted marked differences in the capability of different brands of plastics (identified only by trade names) to cause hemolysis of blood. Further, they found that sterilization with ethylene-oxide gas temporarily increased the hemolytic effects of all three types of tubing. These observations introduce yet another variable, *viz.*, pretreatment, to add to the difficulty of assigning adverse effects to a definite cause. NEERGAARD *et al.* (1971) reported three cases of hepatitis of patients in a hemodialysis unit. The symptoms disappeared rapidly when the plastic tubing was replaced. MCCANDLESS (1971) reported four cases of ileocolitis following exchange transfusion. He considered that prewarming thc blood in a plastic coil could be the source of the trouble.

KISSMEYER-NIELSEN (1961) and KISSMEYER-NIELSEN and MADSEN (1961) examined platelets of blood after storage in untreated glass bottles, silicon-treated glass bottles, and plastic bags. These workers found that platelet number, function, and survival were essentially identical in the three containers. ROLOVIC and BALDINI (1967) found that platelet viability was poor when stored as a concentrate at 4°C in all experiments but that viability was significantly higher in PVC bags. Conversely, clot retraction was found to be less favorable than in glass containers. OZGE *et al.* (1964) found no significant difference in the rate of decrease in platelet number between samples of platelet-rich plasma preserved in plastic and those preserved in siliconized and nonsiliconized glass containers; however, the clot retraction and serotonin-uptake properties of the platelets disappeared more rapidly when the plasma was preserved in plastic. This difference was not found when whole blood was stored at 4°C.

In 1970 MARCEL and NOEL (1970 a and b) and JAEGER and RUBIN (1970 a, b, and c) reported the presence of phthalates in blood stored in PVC bags. MARCEL and NOEL reported that the concentrations of dihexyl phthalate were 0, 4, 7, 11.5, and 11.5 mg/100 ml after storage periods of 0, 4, 8, 15, and 21 days, respectively. GESLER and KARTINOS (1970) pointed out that the material measured by MARCEL and NOEL was almost certainly di(2-ethylhexyl) phthalate (DEHP) and calculated that the amount received even in a complete exchange transfusion would be three orders of magnitude below the toxic level. JAEGER and RUBIN (1970 b) found five to seven mg/100 ml after storage for 21 days at 4°C. RUBIN (1972) stated that while the rate of extraction by blood is 0.25 mg/100 ml/day, no detectable extraction was found with 0.9 percent saline, five percent dextrose, or five percent protein hydrolysates. VALERI *et al.* (1973) gave a figure of 0.32 to 0.5 mg/450 ml/day over a 33-day period of storage at 4°C. The agreement with JAEGER and RUBIN is good, since the increase in surface area in contact with the blood is presumably less in proportion to the volume. SCHETTINO and LA ROTONDA (1969 a) noted some extraction of phthalates with physiological saline and very little with urine. JAEGER and RUBIN (1970 a) showed the presence of both butyl glycolbutyl phthalate and di(2-ethylhexyl) phthalate in blood from plastic storage bags and after circulating through plastic tubing. It was found that butyl glycolbutyl was metabolized by the liver and excreted as glycolyl phthalate, whereas di(2-ethylhexyl) phthalate accumulated in the liver, using a perfused rat liver preparation. JAEGER and RUBIN (1972) measured tissue levels in patients receiving blood and in some subjects who died accidentally and did not receive transfusions. In the latter cases, no phthalates were found in abdominal fat. However, a recent note by RUBIN and NAIR (1973) indicated that 84 out of 140 persons without any history of transfusions were found to have phthalates in their tissues; no levels were given. Although JAEGER and RUBIN (1972) found DEHP in some of the tissues examined (13 to 92 ppm in lung, 270 ppm in one sample of abdominal fat) no clear relationship between the calculated amount of DEHP given in the blood and tissue levels was found. One person received 63 units of blood containing an estimated 600 mg of DEHP, but none was detected in tissues at autopsy. The time interval between transfusion and death would be expected to be a factor, and variation in the rates of absorption and metabolism may also be a factor. The localization of DEHP in blood has been reported by JAEGER and RUBIN (1973). They divided it into three fractions: triple-washed red blood cells and two plasma fractions divided on the basis of density (<1.21 and >1.21). They found that little DEHP occurred in the red cells and that approximately ⅔ was located in the heavy plasma fraction.

Although the extraction of phthalates by blood has been clearly demonstrated, evidence of any physiological effect is lacking. TRIMBLE *et al.* (1966) found that material extracted from Tygon tubing was physiologically active in cold tolerance tests. They identified butyl phthalyl butyl glycolate (BPBG) as a contaminant in brown fat tissue and then found that added BPBG altered the cold tolerance of guinea pigs. The dosage used for these addition experiments was large, 500 to 1,500 mg, and no evidence was presented to show that this amount of BPBG was extractable from the tubing. In other cases, there is some evidence to suggest that materials other than plasticizers are involved. LAWRENCE *et al.* (1963) investigated the tissue responses following intramuscular and subcutaneous implantation of many plastic tubings (primarily PVC) produced by different manufacturers.

It was found that 25 of the 48 samples tested produced toxic responses. Two of these samples were extracted for 12 hours with 75 percent alcohol, rinsed with distilled water, and retested. None of the experiments showed any difference from the original experiment. The extracted material was then separated by gas chromatography into a "light" (less than 16 minutes) and a "heavy" fraction. The "heavy" fraction, which consisted of 95 percent of the extractable material, was phthalates. Nonreactive plastic strips were soaked in either "light" or "heavy" samples and only the "light" soaked strips were found to cause tissue response. Thus, the studies of LAWRENCE *et al.* clearly show that the phthalates were not involved.

GULLBRING *et al.* (1964) studied the inhibition of acetylcholine response, hemolytic effects, and HeLa cell cultures in the guinea pig intestine with eluates of various PVC's. The results were quite variable, indicating differences in chemical composition (unknown) of the PVC's. The inhibition of acetylcholine response was negative for phthalates, although positive for two of the seven PVC's. MORI *et al.* (1967) found that the addition of DEHP did not alter the tissue reaction to cyanacrylates when used for surgical procedures. DUKE and VANE (1968) showed that the pulmonary pressor response to hypoxia was abolished in anaesthetized cats when the blood was circulated through PVC tubing; however, the normal pulmonary vascular responses to a variety of drugs (noradrenaline, 5-hydroxytryptamine, and prostaglandin F_2) were obtained. The two PVC's used had acetyl-tri-*n*-butyl citrate as plasticizers; one was stabilized with octyltin and the other with calcium-zinc. The latter type of tubing was less harmful; both were listed as "surgical non-toxic quality."

IX. Plastics and foods

The extraction of phthalates by milk has been extensively studied by WILDBRETT and co-workers (WILDBRETT *et al.* 1968, 1969, and 1970; WILDBRETT 1973). Two types of PVC were studied, one containing 47.2 percent dinonyl phthalate and the other containing 5.5 percent di(2-ethylhexyl) phthalate and 26.8 percent polyadipate. The migration of phthalic esters was studied in tubing with a length of one m and containing 100 ml of milk. The rate of migration from the plastic containing dinonyl was 4.6 mg/100 ml/day at 38°C and 7.0 mg/100 ml/day at 56°C, while the rate for DEHP was 2.0 mg/100 ml/day at 38°C and 3.1 mg/100 ml/day at 56°C. These rates are higher than those found by RUBIN and JAEGER for blood, but the area of contact would be higher in WILDBRETT's experiments. In additional experiments these workers simulated the actual conditions used, *viz.*, alternate cleaning and milking. They found that the migration rate increased to a maximum at between 90 and 180 hours but had decreased to below the initial rate by 360 hours. Considering the period as a whole, this suggests that some 40 mg of DEHP would be eluated over a 15-day period. No data appear to be available for the levels actually found in milk, but it would be one likely route of exposure of man to phthalates. Nevertheless, since the rate of flow of milk into storage is high, the levels would be expected to be low compared to the values in blood stored in PVC containers; CERBULIS and ARD (1967) did not detect phthalates in commercial milk.

PFAB (1967) placed cheese and lard in contact with two plastic films containing 150 μg of dibutyl phthalate/sq.cm and 270 μg of dicyclohexyl phthalate/sq. cm, respectively. Exposure was for one month at 25°C and a pressure of one to two g/sq. cm was applied to keep the food in contact. It was found that less than one percent of the plasticizers migrated into the food and the concentration of plasticizers in the food was less than two ppm. Levels below three ppm were found by WOGGON and KOCHLER (1967).

X. Metabolism

The most detailed study of the metabolism of phthalates is that of ALBRO *et al.* (1973), who characterized the metabolites by infrared spectroscopy, nuclear magnetic resonance spectroscopy, and mass spectrometry. Their proposed route for the metabolism of di(2-ethylhexyl) phthalate in the rat is given in Figure 1. The first stage is the formation of mono-2-ethylhexyl phthalate, although this compound does not appear in the urine. They found no evidence of conjugation, and the amount of free phthalic acid was less than three percent of the total urinary metabolites. Although the analytical work was less detailed (infrared spectroscopy only), the work of CHAMBON *et al.* (1971) suggests that the metabolites of diethyl and dibutyl phthalates are different from those of DEHP. These workers found that the monoethyl and monobutyl esters were the principal metabolites in the urine of rats. The work of STALLING *et al.* (1973) shows that the metabolites of DEHP in the fish are very different from those formed by the rat. The principal metabolites recovered after 24 hours were free mono-2-ethylhexyl phthalate (66 percent) and conjugated mono-2-ethylhexyl phthalate (14 percent).

SCHULZ and RUBIN (1973) found very rapid disappearance of labeled DEHP from blood when rats were injected intravenously. The plot of log DEHP against time showed a biphasic relationship, the initial slope giving a half-life in the blood of nine minutes and the subsequent slope a half-life of 22 minutes. These workers found that eight percent of the total DEHP injected was converted to water-soluble metabolites within one hour; the metabolites were located primarily in the liver, intestinal contents, and urine. After 24 hours, 54.6 percent was recovered as water-soluble metabolites (primarily in the intestinal contents, excreted feces, and urine) and only 20.5

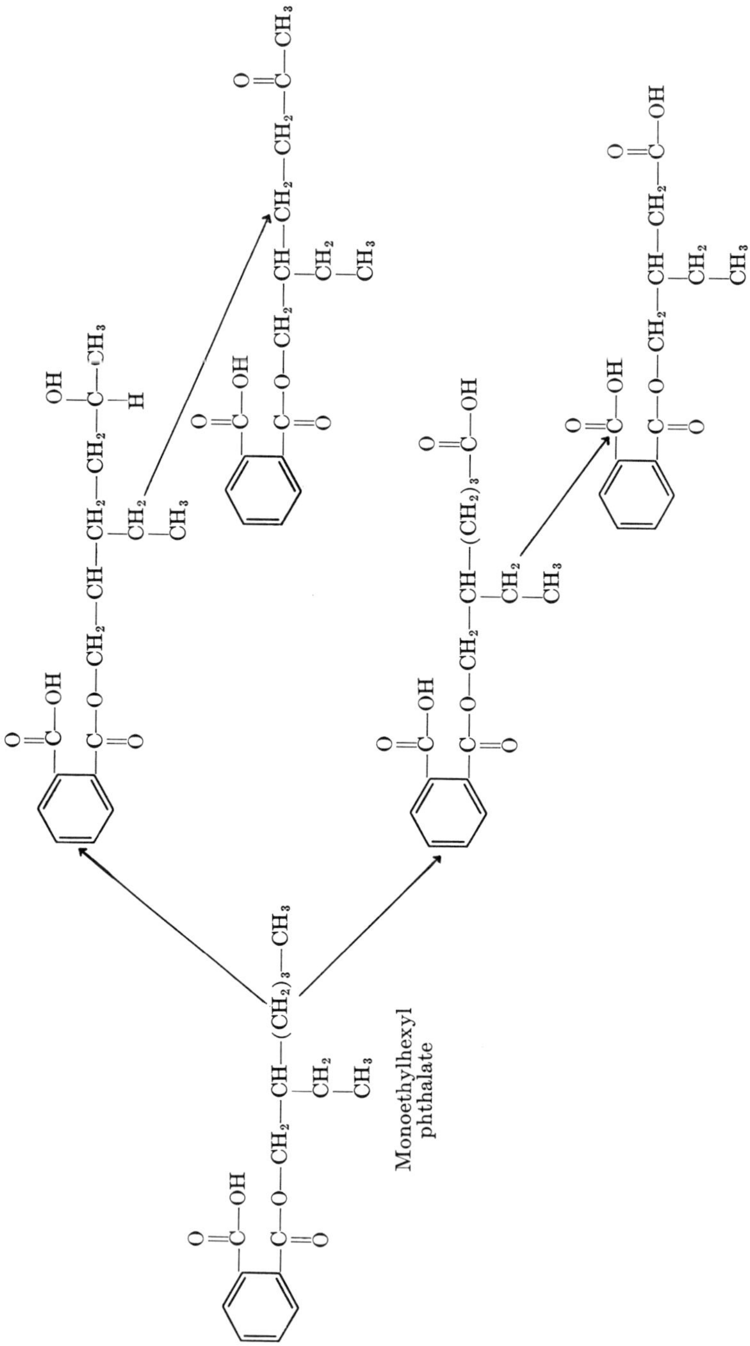

Fig. 1. Routes of metabolism of di(2-ethylhexyl) phthalate (after ALBRO *et al.* 1973).

percent was recovered in the organic extractable form. When DEHP was given orally, 62 percent was recovered as water-soluble metabolites after 24 hours, and 13 percent as organic-soluble material. JAEGER (in JAEGER and RUBIN 1973) gave rats a single i.p. injection of DEHP and examined whole body homogenates 2, 4, and 13 days afterwards. Approximately ¼ of the DEHP was found in the rat after 13 days. A single i.v. injection was given to two rats 24 hours before sacrifice. The total recovery was 79 percent, with 27 percent in the liver (1.6 mg/g) and 27 percent in the lungs (8.2 mg/g). The concentration in the spleen was 1.9 mg/g and was low in all other organs tested. The remaining 21 percent was listed as in carcass. The rather limited data available on mammals indicate rapid metabolism of at least most ingested DEHP.

The rate of metabolism cannot be obtained from the data of STALLING *et al.* (1973), as their fish were continuously exposed to the phthalates. An *in vitro* study using a microsomal preparation of catfish liver showed that dibutyl phthalate was metabolized 16 times more rapidly than diethylhexyl phthalate.

XI. Environmental levels and movement

While an attempt is made to look at environmental movement of phthalates, it must be made clear that the type of detailed information used by WOODWELL *et al.* (1971) for DDT and NISBET and SAROFILM (1972) for polychlorinated biphenyls (PCB's) just does not exist. PAZINSKI (1973) discussed, in general terms, the possible routes of entry of phthalates into the environment and concluded that, "It does not appear that there will be any substantial input or entry from man-made occurrence." While this may be correct, a similar statement on PCB's could just as reasonably have been made a few years ago. Actual levels of phthalates in air appear to be nonexistent and levels in water to be low (HITES and BIEMANN 1972, CORCORAN 1973). HITES and BIEMANN found levels of 0.88 to 1.9×10^{-9} g/g in the Charles River, Boston, with an increase from the mouth of the river towards the more industrial areas of the city; the analysis was confirmed by computerized mass spectrometry/gas chromatography. CORCORAN (1973) found 600×10^{-9} g/g phthalates [assuming all materials to be di(2-ethylhexyl) phthalate], although identification of the phthalates was based on gas chromatography alone. If the effluent of the Mississippi River contained only 0.2 ppm (the author arbitrarily reduced the calculated amount to ⅓), then CORCORAN calculated that 350×10^{6} lb would enter the Gulf of Mexico annually. Even this reduced figure is 40 percent of the total U.S. production (see Table I) and hence the figure of 600×10^{-9} g/g must be due to an abnormal local concentration or be due to an error in the analytical procedure. Using the formula developed by

NISBET and SAROFILM (1973), and considering 1×10^{-9} g/g as the value for a river with heavy industry, then the annual amount of phthalates transported to the oceans by surface run-off is 0.8 million lb.

Amounts of phthalates entering the environment from the various usages are difficult to estimate. The uses listed in Table IV can be divided into four broad categories:

(1) High likelihood of escape. This category includes the nonplasticizer uses (pesticide carriers, cosmetics, fragrances, munitions, oils, and insect repellent). In these cases the phthalate is not encased in a polymer matrix and in some cases is directly added to the environment. Annual U.S. usage is 50×10^6 lb.

(2) Directly in contact with liquids. This amounts annually to 57×10^6 lb as swimming pool liners (20×10^6 lb), garden hoses (15×10^6 lb), and medical uses (22×10^6 lb). While there is considerable evidence that phthalates can be eluated (see sections VIII and IX), the published figures suggest that only a few mg/g of plastic are extracted. A figure of one percent/year would appear to be a maximum for loss to water via direct contact.

(3) Film in contact with air. This amounts annually to 488×10^6 lb as flooring (150×10^6 lb), weather stripping (13×10^6 lb), furniture (90×10^6 lb), auto upholstery and seat covers (80×10^6 lb), wall coverings (38×10^6 lb), auto mats and tops (27×10^6 lb), clothing (72×10^6 lb), and food wrap film (18×10^6 lb). LORANSKII *et al.* (1970) gave levels of 0.15 to 0.26 mg/cu.m phthalates in new rooms with PVC flooring and further stated that phthalates will not be in the air after 1.5 to 2.0 months. It is difficult to calculate the amount of phthalates lost from this information, but if one assumes that the room is two m high and the turn-over time of the air is one hour, then the loss of phthalates is ten mg/sq. cm/day. If we assume that the rate decreases linearly to zero after 50 days, then the total loss of phthalates is 250 mg/sq. cm. Assuming a thickness of 0.25 cm and 30 percent phthalate content, then only $1/3{,}000$ of the original phthalate is lost. POPOV and YABLOCHKIN (1967) measured the concentration of gases released by polyvinyl chloride film (200 μ thick) at 20, 40, and 100°C in a hermetically sealed flask (at atmospheric pressure?). The values in mg/cu.m for dibutyl phthalate were 6.4 (30 days at 20°C), 8.8 (30 days at 40°C), 31.2 (one hour at 100°C), 33.2 (two hours at 100°C), and 35.6 (three hours at 100°C). With one sq. m of film/cu.m of flask, then the loss at 20°C was one part in 300. It would be expected that the rate of loss would be considerably greater for thick flooring as compared to thin film. DVOSKIN *et al.* (1969) studied the loss of dibutyl phthalate from rolls of plastic used in the living quarters of ships. They found that the loss of phthalate from three different samples ranged from 0.53 to 1.06 mg/cu.m (24 hours at 20°C) which increased to 1.30 to 2.60

mg/cu.m at 40°C after 24 hours. Increasing the length of exposure to 120 hours increased these values to 1.06 to 2.13 mg/cu.m (20°C) and 1.80 to 3.73 mg/cu.m (40°C), respectively. Although the data do not allow calculation of the fraction of phthalate lost, the values are of the same order as found by other workers. A figure of 0.1 percent/year (most of which will occur in the first few weeks) is suggested for the rate of loss.

(4) Materials with low surface contact. This amounts annually to 281×10^6 lb as cable and wire (185×10^6 lb), houseware (40×10^6 lb), and miscellaneous (56×10^6 lb).

Quite arbitrarily, a figure of 0.01 percent/year is used. On the basis of these assumptions the total loss to the environment by category is (1) = 50×10^6 lb/year, (2) = 0.57×10^6 lb/year, (3) = 0.49×10^6 lb/year, and (4) = 0.03×10^6 lb/year. Also on the basis of these assumptions the amount eluated directly from plastics while in use is about one million lb annually and is small compared to non-plastic uses.

Two other inputs to the environment need to be considered. The first is loss during manufacture and processing and the second is the ultimate fate of plastic. The losses during use are difficult to get at, but with a volume approaching a billion lb/year, a loss factor of 0.1 percent still involves a million lb of phthalates. Since calculations indicate that only a very small fraction of phthalates in plastics escapes during use, it is obviously of importance to get at the ultimate fate of these materials. NISBET and SAROFILM (1972), considering the disposal of PCB's, estimated that ten to 20 percent was destroyed by high-temperature incineration and two percent was vaporized by open burning in dumps and by small, low-temperature incinerators. The remaining material was disposed of in dumps. The thermal stability of phthalate esters is considerably less than PCB's and even low-temperature incineration or open burning in dumps would be expected to destroy a fair percentage of phthalates. A factor of ⅓ is arbitrarily assigned to cover this decomposition. Although the useful life of plastic objects varies considerably, most will be discarded within a few years. Since the production of phthalates for use in plastics has been reasonably constant at about 900 million lb/year for the last few years in the United States (see Table XIV), the rate of production can be assumed to be equal to the rate of discard.

The major reservoir of plastics is dumps, and thus the question of the rate of loss of phthalates from this reservoir is important. Considerable variation can be expected; the plastic cup in a ditch is more exposed to breakdown than a large piece of plastic in a land-fill operation. VENRICK *et al.* (1973) succumbed to the temptation to extrapolate from the sample of six plastic bottles sighted in the North Pacific Ocean far from the trade routes and came up with a figure 34 million plastic bottles in the North Pacific. CARPENTER and SMITH

Table XIV. *Fate of phthalates used in plastics in the United States.*

Fate	Millions lb/year
Added to dumps	690.0
Destroyed by incineration (15%)	135.0
Vaporized by incineration (2%) (⅓ escaping decomposition)	6.0
Vaporized from plastics in use	0.5
Leached from plastics in use	0.5
Leached from plastics in dumps	?
Losses in handling	1.0

(1972) found that plastic particles average 3,500 pieces and 290 g/sq. km in the western Sargasso Sea. Since the lowest concentrations were found at the end of the Gulf Stream it may well be that the calm area of the Sargasso Sea acts to concentrate this garbage. CARPENTER and SMITH commented that many pieces were brittle, which suggests loss of plasticizers. SCOTT (1972) noted 63 plastic objects in 100 yards and 52 in 50 yards of inaccessible shoreline in Scotland. These were mainly by-products of international commerce, not from local use. Many plastics involved do not contain phthalates, but ocean debris may well be eventually broken down completely.

In terrestrial situations the leaching of phthalates is likely to be very slow unless the plastic is first broken down by other forces. DECOSTE (1968) reported ants eating plasticized wire insulation and a bacteria, *Streptomyces rubrireticuli*, has been reported on PVC material (PAZINSKI 1973). BERK *et al.* (1957), examining the effects of 24 species of fungi, noted that phthalic esters do not serve as carbon sources for fungi. KLAUSMEIER and JONES (1960) studied the metabolism of phthalates in enrichment cultures. They found that di-*n*-octyl and diallyl phthalates did not support growth and that growth with dimethyl phthalate was scanty. Moderate growth was noted with diethyl, dipropyl, dibutyl, butyl isodecyl, *n*-octyl *n*-decyl, and iso-octyl iso-decyl phthalates. Unfortunately, DEHP, the one most widely used, was not tested.

PERRY and SCHELD (1968) reported an *Arthrobacter* isolated from *o*-phthalic acid, and MIYOSHI and HARADA (1970) found a bacterial strain capable of converting 2-butyne-1,4-diol to dibutyl phthalate.

The data do not appear to exist to make a reasonable estimate of the rate of escape of phthalates from dumps. In view of the amount of material involved (ten thousand million lb) such studies would be valuable. Excluding this unknown, the environmental inputs resulting from the usage of phthalates in plastics are estimated at about eight million lb/year, 75 percent of this figure being contributed by low-temperature burning. The losses as vapor may run as high as

90 percent (Table XIV). The risk of loss to the environment of phthalates not used in plastics is clearly much higher. Some uses such as pesticide carriers and insect repellents would enter the environment as part of their normal usage. A detailed breakdown of these nonplastic uses is not available, but it seems likely that one-half might escape to the environment. Thus, these relatively minor usages are probably responsible for more environmental input than the losses from plastic usages. Again, the major loss would be expected to be via vaporization.

METCALF *et al.* (1973) have studied the movement of labeled DEHP in a model ecosystem. The organisms studied were water flea (*Daphnia magna*), mosquito larva (*Culex pipens*), clam (*Sphaerium striatinum*), guppy (*Lebistes reticulatus*), and an aquatic plant (*Elodea canadensis*). The level of DEHP in the water reached a maximum of 0.031 ppm on the fifth day and had decreased to 0.00034 ppm by the end of the 33-day experiment. At that time other levels were algae 18.32, snails 7.30, mosquito larva 36.6, and fish 0.044 ppm. Of the organisms studied, only the guppy showed appreciable rates of metabolism. The DEHP level in the guppy decreased from 88.5 percent of total radioactivity after one day to 37.1 percent after two days and was 16.8 percent after seven days. There was a commensurate increase of polar metabolites. The biomagnification in lower organisms was similar to that found under identical conditions for DDT (METCALF *et al.* 1971). METCALF concluded that "The biomagnification of DEHP together with its teratogenic properties and its enormous rate of production and ubiquitous use indicate the need for much further study of its environmental distribution and fate. Present data suggest the need for restrictions on the use and waste disposal of DEHP." This conclusion should be tempered with the finding that metabolism is fairly rapid in fish and rapid in mammals (see section X). Thus, the largest concentrations would be expected at intermediate points, rather than at the top of the food chain as occurs with DDT. The occurrence of high accumulation factors in aquatic organisms for both di-*n*-butyl and di-(2-ethylhexyl) phthalates has been confirmed by MAYER and SANDERS (1973).

Levels in biological samples have been determined by MAYER *et al.* (1972) and ZITKO (1973). The results of MAYER *et al.* are in Table XV. The results are also expressed as a ratio to the levels of PCB's found, since these are a widespread, stable environmental contaminant for which we have much data. The ratio is above unity in ⅔ of the samples for which the ratio can be calculated. The annual production of DEHP is some ten times that of all PCB's, based on 1970 figures representing the last year of unrestricted use of the PCB's. Even so, the data of MAYER and co-workers suggest that phthalates are widely distributed and possess reasonable stability. ZITKO (1973) found 14.1 ppm (lipid basis) of dibutyl phthalate in an egg of a double-crested cormorant (*Phalacrocorax auritus*) and 10.9 to 19.1

Table XV. *Levels of phthalates and polychlorinated biphenyls in water and fish in North America* (after MAYER *et al.* 1972).

Sample	Site	Di-*n*-butyl phthalate (ng/g)	DNBP/PCB	Di(2-ethylhexyl) phthalate (ng/g)	DEHP/PCB	Polychlorinated biphenyls (ng/g)
Water	Lake Huron, Mich.	0.04	13.0	—	—	trace[a]
		—	—	5.0	10.0	0.2
	Missouri River, Mo.	0.09	0.45	4.9	24.0	0.2
	Lake Superior, Ont.	—	—	300	—	—
Sediment	Lake Superior, Ont.	100	2.5	200	5.0	40
Channel catfish	Miss. and Ark. (industrial areas)	trace	—	3,200	8.0	400
	Iowa (industrial)	200	0.3	400	0.6	700
Walleye	Lake Superior, Ont.	—	—	800	0.6	1,300
Yellow perch	Iowa (industrial)	—	—	—	—	—
Brook trout	California (mountains)	—	—	—	—	—
Dragonfly naiads	Iowa (industrial)	200	1.0	200	1.0	200
Tadpoles	Iowa (industrial)	500	0.5	300	0.3	1,000
Commercial fish food	—	—	—	2,000	2.9	700
	—	—	—	7,000	116	60

[a] Authors reported 0.003 ng/g.

ppm in three eggs of herring gulls (*Larus argentatus*). Two hatchery-reared juvenile Atlantic salmon (*Salmo salar*) had 12.9 and 16.4 ppm of di(2-ethylhexyl) phthalate and the blubber of a common seal pup (*Phoca vitulina*) contained 11 ppm. The collection sites of the eggs and seal were not stated, but were probably from New Brunswick.

Acknowledgments

I am grateful to the National Institutes of Environmental Health Sciences for the translation of eastern European papers. Dr. P. R. Graham of the Monsanto Chemical Company kindly supplied data on the physical properties of the phthalates.

Summary

The annual production of phthalates in the United States is nearly half a million tons and total world production is estimated at two million tons. Phthalates appear to be natural constituents of some plants; evidence for endogenous occurrence in animals is less convincing.

The acute toxicology of all phthalates is low and chronic experiments show a high "no-effect" level. Evidence of carcinogenic properties is negative, but several phthalates are teratogenic at high doses. The risk at environmental levels remains to be assessed.

A considerable body of research has been done on the eluation of phthalates from plastic by blood. No physiological effects have definitely been ascribed to these compounds at these levels.

Accumulation factors of di(2-ethylhexyl) phthalate at the lower trophic levels are high. Since metabolism is rapid in mammals, it is likely that maximum residue levels will occur at intermediate trophic levels rather than the top of the food chain. Current environmental residue data are not adequate to confirm this prediction.

References

AEBI, A., J. ASSELINEAU, and E. LEDERER: Sur les lipides de la souche humaine "Brévannes" de *Mycobacterium tuberculosis.* Bull. Soc. Chim. Biol. **35,** 661 (1953).

ALBRO, P. W., R. THOMAS, and L. FISHBEIN: Metabolism of diethylhexyl phthalate by rats. Isolation and characterization of the urinary metabolites. J. Chromatogr. **76,** 321 (1973).

American Conference of Government Industrial Hygienists: Theshold limit values of airborne contaminants for 1972. Cincinnati (1972).

ASO, K., M. MIGATA, and M. TOMODA: On the production of phthalic acid by Azotobacter. Proc. 2nd Internat. Congress Soil Sci. 3, 40 (1932).

ASSELINEAU, J.: Sur la composition des lipides de *Corynebacterium diphtheriae.* Biochim. Biophys. Acta **54,** 359 (1961).

AUTIAN, J.: Toxicity and health threats of phthalate esters: Review of the literature. Toxicology Information Response Center (1972).

—— Toxicity and health threats of phthalate esters: Review of the literature. Environ. Health Perspect. 4, 3 (1973).

BAROCHE, C.: Actions tératogène et mutagène du phtalate acide de potassium chez la Drosophile. Bull. Cancer **52**, 241 (1965).

—— Actions de la thalidomide chez la Drosophile: Développement et tumorigenese. Comparaison avec le phtalate de potassium et quelques composes non phtaliques. Bull. Cancer **55**, 413 (1968).

BATTA, A. K., and S. RANGASWAMI: Chemical constituents of *Polyporus dryadeus* var. brevisporus and *Fomes allardii*. Phytochem. **8**, 1309 (1969).

BAUMAN, J., R. E. CAMERON, G. KRITCHEVSKY, and G. ROUSER: Detection of phthalate esters as contaminants of lipid extracts from soil samples stored in standard soil bags. Lipids **2**, 85 (1967).

BERK, S., H. EBERT, and L. TEITELL: Utilization of plasticizers and related organic compounds by fungi. Ind. Eng. Chem. **49**, 1115 (1957).

BEROZA, M., B. A. BIERL, and J. M. RUTH: Di-(2-ethylhexyl) phthalate in thermally oxidized corn oil. J. Amer. Oil Chem. Soc. **45**, 441 (1968).

BLICKENSDORFER, P., and L. TEMPLETON: A study of the toxic properties of diethyl phthalate. J. Am. Pharm. Assoc. **19**, 1179 (1930).

BOHRMANN, H., E. STAHL, and H. MITSUHASHI: Studies of the constituents of Umbelliferae plants. 8. Chromatographic studies on the constituents of *Cnidium officinale* Makino. Chem. Pharm. Bull. **15**, 1606, (1967).

BOURNE, H. G., H. T. YEE, and SEFERIAN: The toxicity of rubber additives: Findings from a survey of 140 plants in Ohio. Arch. Environ. Health **16**, 700 (1968).

BOWER, R. K., S. HABERMAN, and P. D. MINTON: Teratogenic effects in the chick embryo caused by esters of phthalic acid. J. Pharmacol. Exp. Ther. **171**, 314 (1970).

BRAUN, B., and H. J. KUMMEL: Uber die Verwendungsmöglichkeit von Kunststoffbehaltern zur Aufbewährung von Blut und Infusionslösungen. Dtsch. Apoth-Ztg. **103**, 467 (1963).

BREGER, I. A.: Geochemistry of lipids. J. Amer. Oil Chem. Soc. **43**, 197 (1966).

—— Phthalates in oil. Chem. Eng. News, p. 19, Jan. 17 (1972).

BREWER, N. R., A. J. CARLSON, and A. PRATT: Chronic toxicity studies on alkyl phenyl phthalate. Food Research **20**, 531 (1955).

BROWN, V. K., C. M. MUIR, and E. THORPE: A contribution to the toxicology of some alcohol mixtures containing 7 to 9 and 9 to 11 carbon atoms and the corresponding phthalate esters. Arch. Toxik. **26**, 84 (1970).

CALLEY, D., J. AUTIAN, and W. L. GUESS: Toxicology of a series of phthalate esters. J. Pharm. Sci. **55**, 158 (1966).

CARPENTER, E. J., and K. L. SMITH, JR.: Plastics on the Sargasso Sea surface. Science **175**, 1240 (1972).

CERBULIS, J., and J. S. ARD: Method for isolation and detection of dioctyl phthalate from milk lipids. J. Assoc. Official Anal. Chemists **50**, 646 (1967).

CHAMBON, P., M. RIOTTE, M. DAUDEN, R. CHAMBON-MOUGENOT, and J. BRINGUIER: Etude du metabolisme des phtalates de dibutyle et de diethyle chez le rat. C. R. Acad. Sc. Paris **273**, 2165 (1971).

CIFRULAK, S. D.: Spectroscopic evidence of phthalates in soil organic matter. Soil Sci. **107**, 63 (1969).

CORCORAN, E. F.: Gas-chromatographic detection of phthalic acid esters. Environ. Health Perspect. 3, 13 (1973).

COVELLO, M., C. RANAUDO, and O. SCHETTINO: Determination of organic phthalates in oils. Rass. Chim. **19**, 236 (1967).

CROSS, B. E., R. H. B. GALT, J. R. HANSON, P. J. CURTIS, J. F. GROVE, and

A. MORRISON: New metabolites of *Gibberella fujikuroi*. II. The isolation of fourteen new metabolites. J. Chem. Soc., pp. 2937–2943 (1963).

DECOSTE, J. B.: Soil burial resistance of vinyl chloride plastics. Ind. Eng. Chem. 7, 238 (1968).

DHINGRA, S. N., D. R. DHINGRA, and G. N. GUPTA: Essential oil of kewda. Perfumery Essent. Oil Record **95**, 219 (1954).

DILLINGHAM, E. O., and J. AUTIAN: Teratogenicity, mutagenicity, and cellular toxicity of phthalate esters. Environ. Health Perspect. **3**, 81 (1972).

DRAIZE, J. H., E. ALVAREK, M. F. WHITESELL, G. WOODARD, E. C. HAGEN, and A. A. NELSON: Toxicological investigations of compounds proposed for use as insect repellents. J. Pharmacol. Exp. Ther. **93**, 26 (1948).

DUEVA, L. A., and M. V. ALDYREVA: Eksperimental'naia otsenka sensibiliziruiushchego i razdrazhaiushchego kozhu deĭstviia ftalatnykh plastifikatorov. Gig. Tr. Prof. Zabol. **13**, 17 (1969).

DUKE, H. N., and J. R. VANE: An adverse effect of polyvinylchloride tubing used in extracorporeal circulation. Lancet **2**, 21 (1968).

DVOSKIN, I. A. G., N. A. RAKHMANINA, F. V. DEM'IANKO, T. A. MEN'SHIKOVA, L. F. EROFEEVA, and A. V. LASHKINA: Hygienic assessment of certain polymers (povinols). [Translated title] Gig. Sanit. **34**, 7 (1969).

ELIZAROVA, O. N.: Comparative toxicologic characteristics of some plasticizers. [Translated title] Mosk. Nauchn.-Issled. Inst. Gigieny No. **9**, 105 (1961).

ETEMADI, A. H.: On the origin of butyl phthalate in bacterial lipid extracts. Bull. Soc. Chim. Biol. **45**, 631 (1963).

FASSETT, D. W.: Organic esters. In F. A. PATTY (ed.): Industrial hygiene and toxicology, vol. 2, pp. 1902–1911. New York: Interscience (1967).

FISHBEIN, L., and P. W. ALBRO: Chromatographic and biological aspects of the phthalate esters. J. Chromatogr. **70**, 365 (1972).

FRANK, O., H. BAKER, S. H. HUTNER, and H. SOBOTKA: Toxicity to protozoa of thalidomide breakdown products—counteraction by nicotinic acid and glutamine. Proc. Soc. Exp. Biol. Med. **114**, 326 (1963).

GAUNT, I. F., J. COOLEY, P. GRASSO, A. B. LANSDOWN, and S. D. GANGOLLI: Acute (rat and mouse) and short-term (rat) toxicity studies on dialkyl 79 phthalate. Food Cosmetic Toxicol. **6**, 609 (1968).

GESLER, R. M., and N. J. KARTINOS: Contamination of blood stored in plastic packs. Lancet **1**, 1227 (1970).

GRAHAM, P. R.: Phthalate ester plasticizers: Why and how they are used. Environ. Health Perspect. **3**, 3 (1973).

GUESS, W. L., S. HABERMAN, D. ROWAN, R. K. BOWER, and J. AUTIAN: Characterization of subtle toxicity of certain plastic components used in the manufacture of the polyvinyls. Amer. J. Hosp. Pharm. **24**, 494 (1967).

GULLBRING, B.: The use of plastics in blood transfusion equipment with special regard to toxicity problems. Vox Sang. **9**, 513 (1964).

——, L. H. EKLUND, and G. SVARTZ-MALMBERG: Clinical and biological test methods applied to plastic products used in transfusion equipment. Vox Sang. **9**, 530 (1964).

HAAGEN-SMIT, F .N. HIROSAWA, and T. H. WANG: Volatile constituents of Zinfandel grapes, *Vitis vinifera*, Var. Zinfandel. Food Res. **14**, 472 (1949).

HABERMAN, S., W. L. GUESS, D. F. ROWAN, R. O. BOWMAN, and R. K. BOWER: Effects of plastics and their additives on human serum proteins, antibodies, and developing chick embryos. Presented Reg. Tech. Conf., Soc. Plastics Eng., N.Y. Sect., pp. 28–43 (1967).

—— —— —— —— —— Effect of plastics and their additives on human serum proteins, antibodies, and developing chick embryos. Soc. Plastics Eng. J. **24**,(3), 62 (1968).

HAYASHI, S., Y. ASAKAWA, T. ISHIDA, and T. MATSUURA: Phthalate esters of

Cryptotaenia canadensis D.C., var. japonica Makino (umbelliferae). Tetrahedron Letters **1967**, 5061 (1967).

Hayashi, T., and T. Nagai: Components of soil humic acids. VIII. Oxidative decomposition with alkaline potassium permanganate and nitric acid. Nippon Dojo-Hiryogaku Zasshi **31**, 197 (1960).

Hemy, G.: The Soviet chemical industry. New York: Barnes and Noble (1971).

Hirose, T., R. Goldstein, and C. P. Bailey: Hemolysis of blood due to exposure to different types of plastic tubing and the influence of ethylene-oxide sterilization. J. Thoracic Cardiovascular Surg. **45**, 245 (1963).

Hites, R. A., and K. Biemann: Water pollution—Organic compounds in the Charles River, Boston. Science **178**, 158 (1972).

Hodge, H. C.: Acute toxicity for rats and mice of 2-ethyl hexanol and 2-ethyl hexyl phthalate. Proc. Soc. Exp. Biol. Med. **53**, 20 (1943).

——, M. R. Goldstein, and Wrightington: Acute toxicity for mice of phthalic acid and certain derivatives. Proc. Soc. Exp. Biol. Med. **99**, 471 (1942).

Homrowski, S. T., and M. Nikonorow: [Title unknown, cited in Piekacz (1971 a)]. Rocz Panstw Zakl Hig. **10**, 321 (1959).

Jaeger, R. J., and R. J. Rubin: Plasticizers from plastic devices extraction, metabolism, and accumulation by biological systems. Science **170**, 460 (1970 a).

—— —— Contamination of blood stored in plastic packs. Lancet **2**, 151 (1970 b).

—— —— Plasticizers from P.V.C. Lancet **2**, 778 (1970 c).

—— —— Migration of a phthalate ester plasticizer from polyvinyl chloride blood bags into stored human blood and its localization in human tissues. N. England J. Med. **287**, 1114 (1972).

—— —— Extraction, localization, and metabolism of di-2-ethylhexyl phthalate from PVC plastic medical devices. Environ. Health Perspect. **3**, 95 (1973).

Japan Plastics Industry Annual, Tokyo: Plastics Age Ltd. (1970).

Jones, W. A., J. D. Warthen, and M. Jacobson: The occurrence of dibutyl phthalate in American cockroaches. Environ. Letters **4**, 103 (1973).

Karel, L., B. H. Landing, and T. S. Harvey: The intraperitoneal toxicity of some glycols, glycol esters, and phthalates in mice. J. Pharmacol. Exp. Therap. **90**, 338 (1947).

Keith, H. B., E. Ginn, G. R. Williams, and G. S. Campbell: Massive hemolysis in extracorporeal circulation. J. Thoracic Cardiovascular Surg. **41**, 404 (1961).

Kemp: Dissertation [Title unknown, cited in Spector (1956)]. Würzburg (1934).

Kissmeyer-Nielsen, F.: Platelets in blood stored in untreated and siliconed glass bottles and plastic bags. I. Studies *in vitro*. J. Clin. Pathol. **14**, 626 (1961).

——, and C. B. Madsen: Platelets in blood stored in untreated and siliconed glass bottles and plastic bags. II. Survival studies. J. Clin. Pathol. **14**, 630 (1961).

Klaus, A., and E. von Sydow: Aroma of cranberries. Acta Chem. Scand. **21**, 2076 (1967).

Klausmeier, R. E., and W. A. Jones: Microbial degradation of plasticizers. Develop. Ind. Microbiol. **2**, 47 (1960).

Knights, B. A.: Isolation of 4-hydroxyphthalide from oat grain. Nature **210**, 1261 (1966).

Kohler, F., W. Meise, and H. Ockenfels: Teratological determination of some thalidomide metabolites. Experientia **27**, 1149 (1971).

Kokurin, A. D., and K. A. Galutkina: Neutral oxygen compounds of heavy tar of bituminous shale. [Translated title] Zh. Prikl. Khim. **40**, 887 (1967).

Kowalski, Z., and E. Bassendowsa: Studies on the acute toxic action of phthalates used in the production of plastics. [Translated title] Med. Pracy. **16**, 109 (1965).

KRAUSKOPF, L. G.: Studies on the toxicity of phthalates via ingestion. Environ. Health Perspect. 3, 61 (1973).

LAWRENCE, W. H., J. L. MITCHELL, W. L. GUESS, and J. AUTIAN: Toxicity of plastics used in medical practice. I. Investigation of tissue responses in animals by certain unit packaged polyvinyl chloride administration devices. J. Pharm. Sci. 52, 958 (1963).

LEHMAN, A. J.: Insect repellants. Assoc. Food and Drug Officials U.S. Quart. Bull. 19, 87 (1955).

LEVINSKAS, G. J.: Inhalation toxicity tests. In: Flexible vinyls and human safety, Soc. of Plastic Eng., pp. 95–100 (1973).

LIVINGSTON, J. R.: (1971). Unpublished work, cited in KRAUSKOPF (1973).

LORANSKII, D. N., S. N. SMIRNITSKII, and E. N. SHAKLEINA: Printsipy gigienicheskogo normirovaniia pri otsenke polimernykh stroietl'nykh materialov. Gig. Sanit. 35, 51 (1970).

MALLETTE, F. S., and E. VON HAAM: The toxicity and skin effects of compounds used in the rubber and plastics industries. II. Plasticizers. Arch. Ind. Hgy. Occup. Med. 6, 231 (1952).

MARCEL, Y. L., and S. P. NOEL: Contamination of blood stored in plastic packs. Lancet 1, 35 (1970 a).

—— —— A plasticizer in lipid extracts of human blood. Chem. Phys. Lipids 4, 418 (1970 b).

MASLENKO, A. A.: Normalization of the dibutyl phthalate level in reservoir waters. [Translated title] Gig. Naselennykh Mest., pp. 72–75 (1967).

—— Hygienic characteristics of dibutylphthalate and dioctylphthalate water pollutants. [Translated title] Gig. Sanit. 33, 102 (1968).

—— Effect of dioctyl phthalate on the organoleptic properties of water and the total sanitary system of reservoirs. [Translated title] Gig. Naselennykh Mest. 8, 80 (1969).

MATSUDA, K., and M. SCHNITZER: Reactions between fulvic acid, a soil humic material and dialkyl phthalates. Bull. Environ. Contam. Toxicol. 6, 200 (1971).

MAYER, F. L., JR., and H. O. SANDER: Toxicology of phthalic esters on aquatic organisms. Environ. Health Perspect. 3, 153 (1973).

——, D. L. STALLING, and J. L. JOHNSON: Phthalate esters as environmental contaminants. Nature 238, 411 (1972).

MCCANDLESS, A. E.: Ileocolitis following exchange transfusion. Lancet 2, 762 (1971).

MCLAUGHLIN, J., JR., J. P. MARLAIC, M. J. VERRETT, M. K. MUTCHER, and O. G. FITZHUGH: The injection of chemicals into the yolk sac of fertile eggs prior to incubation as a toxicity test. Toxicol. Applied Pharmacol. 5, 760 (1963).

MCOMIE, W. A.: Certain aspects of the toxicity of diallyl phthalates. Fed. Proc. 5, 191 (1946).

—— (1949). Title unknown, cited in SPECTOR (1956).

——, and H. H. ANDERSON: Toxicity of diallyl phthalate. Univ. Calif. Publ. Pharmacol. 2, 205 (1947).

MEIJLER, F. L., and D. DURRER: The influence of poly-vinyl-chloride tubing on the isolated perfused rat's heart. Vox Sang. 4, 239 (1959).

MEN'SHIKOVA, T. A.: Hygienic evaluation of dibutylphthalate in relation to the use of polymeric materials for finishing living quarters on ships. [Translated title] Gig. Sanit. 36, 23 (1971).

METCALF, R. L., G. M. BOOTH, C. K. CARTER, D. J. HANSEN, and P. Y. LU: Uptake and fate of di-2-ethylhexyl phthalate in acquatic organisms and in a model ecosystem. Environ. Health Perspect. 4, 27 (1973).

MEYLER, F. L., A. F. WILLEBRAND, and D. DURRER: The influence of poly-vinyl-chloride (P.V.C.) tubing on the isolated perfused rat's heart. Circ. Res. **8**, 44 (1960).

MILKOV, L. E., M. V. ALDYREVA, T. B. POPOVA, K. A. LOPUKHOVA, I. U. L. MAKARENKO, L. M. MALIAR, and T. K. SHAKHOVA: The status of health of workers subjected to the effects of phthalate plasticizers in the production of artificial leather and film (with PVC base). [Translated title] Gig. Tr. Prof. Zabol. **13**, 14 (1969).

MITSUHASHI, H., T. MURAMATSU, U. NAGAI, T. NAKANO, and K. UENO: Studies on the constituents of Umbelliferae plants. VIII. Distribution of alkylphthalides in Umbelliferae plants. Chem. Pharm. Bull. **11**, 1317 (1963).

MIYOSHI, T., and T. HARADA: Formation of dibutylphthalate from 2-butyne-1, 4-diol by a soil bacterium. Biochim. Biophys. Acta **222**, 684 (1970).

MOKHANACHEV, I. G., and L. G. AATAKHOVA: Water soluble di- and tricarboxylic acids in tobacco smoke. Tabak (Moscow) **29**(3), 31 (1968).

MORI, S., K. OTA, M. TAKADA, and T. IONU: Comparative studies of cyanoacrylate derivative *in vivo*. J. Biomed. Material Res. **1**, 55 (1967).

MORRIS, R. J.: Phthalic acid in the deep sea jellyfish *Atolla*. Nature **225**, 1264 (1970).

NAZIR, D. J., M. BEROZA, and P. P. NAIR: Isolation and identification of di(2-ethylhexyl) phthalate—a component of heart muscle mitochondria. Fed. Proc. **26**, 412 (1967).

——, A. P. ALCARAZ, B. A. BIERL, M. BEROZA, and P. P. NAIR: Isolation, identification, and specific localization of di-2-ethylhexyl phthalate in bovine heart muscle mitochondria. Biochem. **10**, 4228 (1971).

NEERGAARD, J., B. NIELSEN, V. FAURBY, D. H. CHRISTENSEN, and O. F. NIELSEN: Plasticizers in P.V.C. and the occurrence of hepatitis in a haemodialysis unit: A preliminary communication. Scand. J. Urol. Nephrol. **5**, 141 (1971).

NEMATOLLAHI, J., W. L. GUESS, and J. AUTIAN: Plasticizers in medical application. I. Analysis and toxicity evaluation of dialkyl benzenedicarboxylates. J. Pharm. Sci. **56**, 1446 (1967).

NISBET, I. C. T., and A. F. SAROFILM: Rates and routes of transport of PCB's in the environment. Environ. Health Perspect. **1**, 21 (1972).

OGNER, G., and M. SCHNITZER: Humic substances—fulvic acid-dialkyl phthalate complexes and their role in pollution. Science **170**, 317 (1970).

OPPENHEIMER, B. S., E. T. OPPENHEIMER, A. P. STOUT, and I. DANISHEFSKY: Malignant tumors resulting from embedding plastic in rodents. Science **118**, 305 (1953).

—— ——, I. DANISHEFSKY, A P. SCOTT, and F. R. EIRICH: Further studies of polymers as carcinogenic agents in animals. Cancer Res. **15**, 333 (1955).

—— ——, A. P. STOUT, I. DANISHEFSKY, and M. WILLHITE: The mechanism of carcinogenesis by plastic films. Acta Unio. Internat. Contra. Cancrum **15**, 659 (1959).

OZGE, A., M. BALDINI, and R. GOLDSTEIN: Effect of plastic and glass surface on clot retraction and serotonin uptake of platelet-rich plasma stored at 4°C. J. Lab. Clin. Med. **63**, 378 (1964).

PARODI, P. W., and R. J. DUNSTAN: Unsaponifiable fraction of butterfat: Contamination with phthalic acid esters. Aust. J. Dairy Technol. **23**(1), 20 (1968).

PATTY, F. A. (ed.): Industrial hygiene and toxicology, vol. 2. New York: Interscience (1967).

PAZINSKI, H. J.: Phthalates in the environment. In: Flexible vinyls and human safety, Soc. of Plastic Eng. (1973).

PERKINS, E. G.: Formation of nonvolatile decomposition products in heated fats and oils. Food Technol. **21**, 611 (1967 a).

—— Characterization of the nonvolatile compounds formed during the thermal oxidation of corn oil. II. Phthalate esters. J. Amer. Oil Chem. Soc. **44**, 197 (1967 b).

PERRY, J. J., and H. W. SCHELD: Oxidation of hydrocarbons by microorganisms isolated from soil. Can. J. Microbiol. **14**, 403 (1968).
PFAB, W.: Migration of phthalate plasticizers from lacquered aluminum foils on fatty foods. [Translated title] Deut. Lebensm.-Rundsch. **63**(3), 72 (1967).
PFUDERER, P., and A. A. FRANCIS: The effect of phthalate esters on the heartrate of the goldfish. Fed. Proc. **32**, 74 (1973).
PHILLIPS, H. F., and I. A. BREGER: Isolation and identification of an ester from a crude oil. Geochim. Cosmochim. Acta **15**, 51 (1958).
PIEKACZ, H.: Effect of dibutyl phthalate and dioctyl phthalate on rat's organism administered orally over a prolonged period. I. Use and toxic properties of dibutyl phthalate and dioctyl phthalate. [Translated title] Rocz Panstw Zakl Hig. **22**, 55 (1971 a).
—— Effect of dioctyl- and dibutylphthalates on rats during oral administration in prolonged experiments. II. Studies of subacute and chronic toxicity. [Translated title] Rocz Panstw Zakl Hig. **22**, 295 (1971 b).
—— Influence on rat organism of dibutyl phthalate and dioctyl phthalate administered orally for a protracted time. III. Influence on fertility and fetus development. [Translated title] Rocz Panstw Zakl Hig. **22**, 519 (1971 c).
POPOV, L. A., and U. D. YABLOCHKIN: Characteristics of gases released by polyvinyl chloride film. [Translated title] Gig. Sanit. **32**, 114 (1967).
PROENCA DA CUNHA, A.: Analytical study of the essential oil of *Eucalyptus punctata* of Angola. [Translated title] Garcia Orta **14**, 411 (1966).
PROTSENKO, E. I.: Gonadotropic action of phthalic anhydride. [Translated title] Gig. Sanit. **35**, 105 (1970).
RADEVA, M., and S. DINOEVA: Toxicity of dibutyl phthalate by oral application in albino rats. [Translated title] Khig. Zdraveopazvane **9**, 510 (1966).
RALL, D. P.: Risks, research, and reason. Fed. Proc. **32**, 1766 (1973).
RANAUDO, C., and O. SCHETTINO: Colorimetric determination of phthalates in olive and olive husk oils. Rassegna Chimica **21**, 171 (1969).
ROLOVIC, Z., and M. BALDINI: Comparative studies of the viability of human platelets stored at 4°C in plastic and glass containers. Transfusion **7**, 204 (1967).
RUBIN, R. J.: Storage of aqueous solutions for parenteral infusion. Lancet **1**, 965 (1972).
——, and P. P. NAIR: Plasticizers in human tissues. N. England J. Med. **288**, 915 (1973).
SCHETTINO, O., and M. I. LA ROTONDA: Release of additives from surgical instruments and pharmaceutical containers made of poly(vinyl chloride) (PVC). I. Chromatographic detection of alkyl citrate and phthalate plasticizers. Boll. Soc. Ital. Biol. Sper. **45**, 1337 (1969).
SCHINZ, H., and C. F. SEIDEL: Untersuchungen über Aromastoffe. I. Über das Himbeeraroma. Helv. chim. Acta **40**, 1839 (1957).
SCHMELTZ, I., and W. S. SCHLOTZHAUER: Nonvolatile acids of cigar smoke. Tobacco Sci. **6**, 88 (1962); Tobacco **154**(20), 32 (1962).
SCHMID, I. H., and P. KARRER: Über wasserlösliche Inhaltsstoffe von *Papaver somniferum L.* Helv. chim. Acta **28**, 722 (1945).
SCHNITZER, M., and J. G. DESJARDINS: The alkaline permanganate oxidation of organic matter extracted from a podzol Bh horizon. Can. J. Soil Sci. **44**, 272 (1964).
SCHULZ, C. O., and R. J. RUBIN: Distribution, metabolism, and excretion of di-2-ethylhexyl phthalate in the rat. Environ. Health Perspect. **3**, 123 (1973).
SCOTT, G.: Plastics packaging and coastal pollution. Internat. J. Environ. Studies **3**, 35 (1972).
SHAFFER, C. B. M., C. P. CARPENTER, and H. F. SMYTH: Acute and subacute toxicity of di(2-ethylhexyl)phthalate with note upon its metabolism. J. Ind. Hyg. Toxicol. **27**, 130 (1945).

SIMS, R.: Analysis of hepatocarcinogenic fish meal lipids. U.S. Fish and Wildlife Service, Res. Rept. No. **70**, 178 (1967).

SINGH, A. R., W. H. LAWRENCE, and J. AUTIAN: Teratogenicity of phthalate esters in rats. J. Pharm. Sci. **61**, 51 (1972).

SMITH, C. C.: Toxicity of butyl stearate, dibutyl sebacate, dibutyl phthalate and methoxyethyl oleate. Arch. Ind. Hyg. **7**, 310 (1953).

SMITH, O. M.: Toxic properties of diethylphthalate. J. Amer. Pharm. Assoc. **13**, 812 (1924).

SMITH, R. L., S. FABRO, H. SCHUMACHER, and R. T. WILLIAMS: The relation between the chemical structure and embryotoxic activity of thalidomide and related compounds. Symp. Embryopathic Activ. Drugs. Univ. Col. London, pp. 194–209 (1965).

SMYTH, H. F., JR., C. P. CARPENTER, C. S. WEIL, and U. C. POZZANI: Range-finding toxicity data: List. Arch. Ind. Hyg. **10**, 61 (1954).

SPASOVSKI, M.: The maximum allowable concentration of dibutyl phthalate. [Translated title] Khigiena **7**, 38 (1964).

SPECTOR, W. S. (ed.): Handbook of toxicology. New York: Saunders (1956).

STALLING, D. L., J. W. HOGAN, and J. L. JOHNSON: Phthalate ester residues: Their metabolism and analysis in fish. Environ. Health Perspect. **3**, 159 (1973).

STEVENS, K. L., J. BOMBEN, A. LEE, and W. H. MCFADDEN: Volatiles from grapes. Muscat of Alexandria. J. Agr. Food Chem. **14**, 249 (1966).

STEWART, J. W., and M. F. STURRIDGE: Haemolysis caused by tubing in extracorporeal circulation. Lancet **1**, 340 (1959).

SUGIYAMA, N., C. KASHIMA, M. YAMAMOTO, T. SUGAYA, and R. MOHRI: Altenin. I. The isolation of Altenin from *Alternaria kikuchiana*. Bull. Chem. Soc. Japan **39**, 1573 (1966).

SWAIN, A. P., W. RUSANIWSKJ, and R. L. STEDMAN: Hexane-soluble substances of tobacco leaves. Chem. & Ind., pp. 435–436 (1961).

TABORSKY, R. G.: Isolation studies on a lipoidal portion of the bovine pineal gland. J. Agr. Food Chem. **15**, 1073 (1967).

THIELE, O. W., and H. G. TRUPER: Esters of phthalic acid as artifacts in bacterial lipids. Arch. Mikrobiol. **2**, 91 (1972).

TRIMBLE, A. S., B. S. GOLDMAN, J. K. YAO, L. K. KOVATS, and W. G. BIGELOW: Plastics—a source of chemical contamination in surgical research. Surgery **59**, 857 (1966).

TYSON, C. A.: (1972). Unpublished work, cited by KRAUSKOPF (1973).

VALERI, C. R., T. J. CONTRERAS, H. FEINGOLD, R. H. SHELBLEY, and R. J. JAEGER: Accumulation of di-2-ethylhexyl phthalate (DEHP) in whole blood, platelet concentrates and platelet-poor plasma. Environ. Health Perspect. **3**, 103 (1973).

VENRICK, E. L., T. W. BACKMAN, W. C. BARTRAM, C. L. PLATT, M. S. THORNHILL, and R. E. YATES: Man-made objects on the surface of the central North Pacific Ocean. Nature **241**, 271 (1973).

VERRETT, M. J., M. K. MUTCHLER, W. F. SCOTT, E. F. REYNALDO, and J. MCLAUGHLIN: Teratogenic effects of captan and related compounds in the developing chicken embryo. Ann. N.Y. Acad. Sci. **160**, 334 (1969).

WAKAYAMA, S., and S. NAMBA: The constituents of the flower oil of lily of the valley. Bull. Chem. Soc. Japan **38**, 860 (1965).

WALTER, C. W.: A technique for collection, storage, and administration of unadulterated whole blood. In: Surgical forum 1950, Amer. Col. Surgeons. Philadelphia: Saunders (1951).

WILDBRETT, G.: Diffusion of phthalic acid esters from PVC milk tubing. Environ. Health Perspect. **3**, 29 (1973).

——, K. W. EVERS, and F. KIERMEIER: Application of plastics in the food industry. XI. Detection of monomeric plasticizers contained in plastics in milk. Z. Lebensm.-Unters. Forsh. **137**, 356 (1968).

—— —— —— Application of plastics in foods. VII. Migration of monomeric plasticizers from PVC milk tubing into cleansing solutions. Fette, Seifen, Anstrichm. **71**, 330 (1969).

—— —— —— Untersuchungen und Betrachtungen zur Anwendung von Kunststoffen für Lebensmittel. XIII. Zum Verhalten von Kunststoffen gegenüber Milch. Z. Lebensm.-Unters. Forsch. **12**, 205 (1970).

WILLIAMS, R. T.: Thalidomide: A study of biochemical teratology. Arch. Environ. Health **16**, 493 (1968).

WOGGON, H., and U. KOEHLER: The testing of plastics consumer goods. Polarographic determination of bis(2-ethylhexyl) phthalate and study of its migration tendency. Kunststoffe **57**, 583 (1967).

WOODWELL, G. M., P. P. CRAIG, and H. A. JOHNSON: DDT in the biosphere: Where does it go? Science **174**, 1101 (1971).

YAMAMOTO, A., and T. OMORI: The flavors in sake. VIII. Separation and identification of phthalic acid ester, *p*-hydroxybenzoic acid, and its ester. Nippon Nogei Kagaku Kaishi 38, 231 (1964).

ZITKO, V.: Determination of phthalates in biological samples. Internat. J. Environ. Anal. Chem. **2**, 241 (1973).

Manuscript received October 15, 1973; accepted January 17, 1974.

Reduction of pesticide residues in food crops by processing

By

J. R. GEISMAN*

Contents

I. Introduction

Agricultural productivity has been on the increase since the 1920's and will likely continue to increase. Much of this increased yield is due to improved control of plant pests. Control of plant pests has become possible through the proper use of pesticides. It seems likely that control measures will continue to be improved with increasing specificity of pesticides. Concomitantly, the quality of food commodities has also improved which may be directly associated with improved human health.

An increasing number of people are becoming concerned about possible health hazards as a result of using pesticides. The legal limitations of pesticide application and usage are becoming more restrictive. However, the possibility of pesticide residues in foods creates a serious emotional situation in society. The U.S. Food and Drug Administration (FDA) has a constant surveillance program and conducts extensive

* Department of Horticulture, The Ohio State University and the Ohio Agricultural Research and Development Center, Columbus, Ohio 43210.

research studies on pesticide use. Laws and regulations governing uses of pesticide chemicals form a rather complex system in the United States.

Processed foods play an important role in the human diet. They have a nutritional input by providing seasonal commodities on a year-round basis and offer a wide variety of tastes for every palate in addition to being economical.

Most fruits and vegetables for processing are subjected to different unit operations during preparation and preservation. These operations are essential for cleaning, eliminating wastes, and making raw materials more palatable. The procedures used may actually reduce or remove pesticide residues, when present.

II. Pesticide residues in foods

DURHAM (1963) reviewed the problem of pesticide residues in foods and the possibility of health hazards. Although there is concern about pesticides in relation to public health, METCALF (1965) reported that since 1947, there have been no records of illness or disease resulting from the proper use of agricultural chemicals. WESSEL (1971), in reviewing the status of residues in foods, noted that from 1964 to 1969 the FDA had sampled 95,736 samples of foods over half of which contained a pesticide residue of one kind or another. The residues found in these samples were extremely low, generally below tolerance.

Of all the various types of pesticides in use, the main residue problem has been from insecticides. WESSEL's (1971) data indicated that from 1967 to 1969 insecticides comprised an average of 89.4 percent of the pesticide residues found in the diet. In this same period fungicides accounted for 6.2 percent of the residues while herbicides were found to contribute only 4.4 percent. Furthermore, the organochlorine compounds were conspicuously present as residues, accounting for 85 percent of the total insecticide residue. Of course, this does represent a picture of the chemical classes of insecticides in use at that time and is in agreement with the data of LISKA and STADELMAN (1969).

III. Unit operations involved in processing

Most commodities for processing are subjected to a number of unit operations depending on both the commodity and the finished product. Specific unit operations that may affect pesticide residues include inspecting; washing; blanching, if applicable; peeling, if applicable; and retorting or pasteurizing.

Inspection of the raw product with subsequent removal of damaged material could reduce residues. Imperfections and convolutions increase surface area, thereby increasing the amounts of surface residues present. These imperfections also make it more difficult to remove residues (GEISMAN and GOULD 1970, FARROW *et al.* 1969 a).

The removal of pesticide residues by any unit operation is dependent on numerous vectors. The nature of the commodity being processed is of prime consideration since this determines which operations are utilized. The pesticide(s) used should also be given consideration as to chemical properties, formulation applied, method of application, and rate of application. Finally, the interaction of the pesticide with the commodity should be kept in mind from the viewpoint of the time period of contact between the chemical and the commodity (FARROW *et al.*, 1969 b).

IV. Effect of processing operations on pesticide residues

The literature indicates that until 1947, little was published on the effects of preparation and processing in removing pesticide residues. One of the first papers to indicate that processing altered the residue was by TRESSLER (1947) who indicated that decomposition of DDT occurred when it was processed with foods. It should be pointed out that prior to this time a general recommendation existed for washing apples in a dilute hydrochloric solution to remove arsenic and other inorganic residues (SMITH *et al.* 1934).

Much research reported between 1948 and 1950 was specifically concerned with the effect of washing and processing on residues of DDT in various commodities. (TRESSLER 1947, LAMB *et al.* 1948 and 1950, WALKER 1949, BOHM *et al.* 1950, BRITTIN and FAIRING 1950, HALLER and CARTER 1950). While evidence was given of reduction of DDT residues due to washing and processing, residue detection was based on colorimetric methods which could not separate and identify all of the isomers or metabolites which are readily obtainable with chromatographic techniques. Evidently, the status of detection methodology caused considerable delay in further studies on the effect of processing on residues until the late 1950's. Then several articles appeared which were more on the role of specific unit operations in removing or reducing pesticide residues.

a) *The washing operation*

Washing including rinsing is, perhaps, the one unit operation common to the preparation of nearly all fruits and vegetables for processing. Various physical and chemical parameters of the operation have only recently been established.

A survey of industry practices revealed that while washing should be done, the specific methodologies were left to the discretion of the individual processor. Investigations were undertaken at the Ohio

Agricultural Research and Development Center (GOULD *et al.* 1959) and at the National Canners Association Research Laboratory (MERCER 1960) to determine the physical and chemical parameters of the washing operation. The physical aspects included soaking time, soaking temperature, agitation during soaking, rotation of commodities under spray rinse, number and type of nozzles, and spray rinse pressure and volume.

The role of commodity in determining the physical aspects of washing is readily apparent from the data in Table I. The physical parameters could be manipulated to produce optimum removal of soil and residues from different commodities. In addition, some commodities may be exposed to multiple washings.

The chemical aspects of washing were wetting agent type and concentration and chlorine concentration. The type of wetting agent was probably most important due to the foaming characteristic. The amount of foam was inversely related to cleaning efficiency. GEISMAN and GOULD (1970) studied tritiated detergents of high, medium, and low foaming ability and reported that high and medium foaming detergents should not be used in washing fruits and vegetables. The reasons for this recommendation were the foam cap interfered with both cleansing and rinsing to the extent that detergent residues could become a contaminant in the finished product (Table II). Also, the loss of detergent due to foam overflow from the soak tank would be excessively expensive. MERCER (1962) made similar recommendations for washing asparagus.

Extraneous matter, if present, could be removed by washing. This information served as a stimulus to investigate the possibility of removing pesticide residues by washing. LAMB *et al.* (1968 a), investigating DDT residues in potatoes, reported a 20 percent reduction in residues by washing. ELKINS *et al.* (1968) indicated that cold water washing removed 96 percent of the malathion residues present in

Table I. *Physical parameters recommended for washing sweet corn* (GEISMAN and GOULD, 1963) *and tomatoes* (GOULD *et al.* 1959).

Physical parameter	Recommendation for	
	Sweet corn	Tomatoes
Soak time	3 min.	3 min.
Soak temperature	100°F	130°F
Agitation	Vigorous	Vigorous
Rotation under rinse	3 revolutions	2 revolutions
Number of nozzles	1/sq.ft.	1/sq.ft.
Type of nozzles	Fullcone; knife	Fullcone
Rinsing pressure	150 psi	130 psi

Table II. *Percent reduction in specific activity of soak water for sound and cracked tomatoes by detergent foam characteristics* (GEISMAN and GOULD 1970).

Foam characteristic	Reduction (%)	
	Sound fruit	Cracked fruit
High	45	64
Medium	38	41
Low	4	5

green beans. LAMB *et al.* (1968 b) reported that minimum commercial washing removed 17 percent of the DDT residues and 66 percent of the carbaryl residues in spinach but did not remove any parathion. However, maximum commercial washing reduced the DDT residues by 45 percent, the carbaryl residues by 87 percent, and the parathion residues by nine percent. The addition of detergent to the wash water increased the amount removed.

GEISMAN and DEPPIN (1967) found that DDT could be completely removed from spinach by washing, if the crop were harvested within one day after application. On the other hand, only 62 percent of the DDT residues could be removed one day after application. They found that as time after application increased, it became more difficult to remove residues by washing, particularly if the pesticide had a persistent residue. GEISMAN and GOULD (1970) gave data on the difficulty of removing DDT residues from tomatoes as a function of time after application (Table III).

Less persistent residues of dieldrin (GEISMAN and GOULD 1970) and malathion (YAO and GEISMAN 1972) could be readily removed by washing regardless of the interval between application and washing.

Table III. *Average percent reduction of DDT-^{14}C from tomatoes by washing at various days after application* (GEISMAN and GOULD 1970).

Days after application	Av. reduction (%)
1	62.0
3	79.5
7	78.0
14	77.5
21	85.0
28	92.5
34	94.0

KOIVISTOINEN and KARINPAA (1965) reported similar results on tomatoes, apples, and plums dipped in IPC and CIPC. KOIVISTOINEN *et al.* (1965) indicated that captan residues could be easily removed by washing strawberries, gooseberries, string beans, tomatoes, plums, and apples in running water.

A recent innovation in the washing operation to conserve water has been suggested by KROCHTA *et al.* (1973). It involves the generation of foam and utilizes mechanical wiping devices to remove nearly all the soil found on tomatoes. It would be worthwhile investigating the effect of this treatment on removing residues.

b) The blanching operation

Blanching is a mild heat treatment or partial cooking usually employed with vegetables. The blanching operation is usually accomplished either in steam or hot water. This operation may also accomplish some washing of the product.

ELKINS *et al.* (1968) investigated both water and steam blanching of green beans. They reported that water blanching removed 50 percent of the DDT residues and 68 to 73 percent of the carbaryl residues. Steam blanching was ineffective in removing these residues. LAMB *et al.* (1968 b) found that water blanching of spinach removed 38 to 60 percent of the DDT, 49 to 71 percent of the parathion, and 96 to 97 percent of the carbaryl. With these materials, steam blanching removed little or no residue. YAO and GEISMAN (1972) found that steam blanching of spinach removed only two to three percent of the malathion residue. FARROW *et al.* (1969 a and b) reported that steam blanching had no effect on parathion residues on broccoli. KOIVISTOINEN *et al.* (1965) suggested that losses of captan were over 90 percent if the commodity were blanched.

FARROW *et al.* (1969 b) inferred that blanching alone could remove a large portion of the residue present. However, when blanching followed washing, an even greater amount of residue was removed indicating a cumulative effect (Table IV).

There are currently some improvements in the methods of blanch-

Table IV. *Removal of pesticide residues from spinach by blanching and washing plus blanching* (FARROW *et al.* 1969 b).

Residue	Removal (%) by	
	Blanching	Washing + blanching
DDT	48	60
Carbaryl	84	97
Parathion	61	71

ing which include individual quick blanching (IQB) (LAZAR *et al.* 1971), microwave blanching (CHEN *et al.* 1971), and hot air blanching (RALLS *et al.* 1972). These methods have not been evaluated for residue removal.

c) *Peeling or trimming operations*

Peeling, when applicable, would remove any surface contaminants. The main disadvantage is that not all commodities can be peeled. Peeling may be done by hand, mechanically, or chemically. Trimming is usually done by hand with a knife usually designed for the commodity. Trimming may also be done mechanically utilizing rotating knife blades designed for the commodity.

LAMB *et al.* (1968 a) studied the effects of chemical peeling and hand peeling on DDT residues in potatoes. They noted that lye peeling removed 74 percent of the residues and hand peeling removed more than 91 percent. It should also be noted that the amount of peel removed by hand was greater than the amount removed chemically. FARROW *et al.* (1968) indicated that peeling of tomatoes had an effect on the residues of DDT, malathion, and carbaryl, removing virtually all of each residue.

FARROW *et al.* (1969 a) stated that a slight decrease in the residue levels of parathion and carbaryl might result if broccoli stalks were trimmed. GEISMAN (1972) has also indicated that trimming could affect residue contents of tomatoes.

A refinement of the peeling operation is the use of dry caustic (RALLS *et al.* 1972). This technique offers reduced wastes and efficient peeling but has not been evaluated for reducing residues.

d) *Heat processing and home preparation*

The treatment of commodities with enough heat to pasteurize or preserve the product may be accomplished in numerous ways. A wide variety of equipment is in use for this purpose. Retorting or pasteurizing is the heat treatment given to preserve the finished product. The main reaction of interest would be degradation of residues, if present, by heat. This reaction could also take place in other unit operations involving heat which have not been mentioned, such as drying or dehydrating. A similar reaction might occur in home cooking of food products as far as effects on residues are concerned.

CARTER *et al.* (1948) were among the first to publish data concerned with the effects of cooking on pesticide residues. They reported that cooking had little effect on DDT residues in beef.

FARROW *et al.* (1966) indicated that DDT was converted to TDE during retorting of canned spinach. TDE was further decomposed during storage. KOIVISTOINEN and KARINPAA (1965) reported that IPC and CIPC residues on tomato, apple, and plum fruits were reduced about 50 percent in the preservation process. FARROW *et al.* (1968) stated that commercial canning and juicing operations removed virtually all DDT, malathion, and carbaryl residues. Similar results were obtained by home canning techniques. Using the same chemicals on green beans, ELKINS *et al.* (1968) found DDT was converted to TDE during heat processing. Commercial processing removed 94 percent of the malathion and home cooking left only trace amounts. Carbaryl residues seemed to be unaffected by heat. CARLIN *et al.* (1966) obtained similar results with azinphosmethyl and DDT on snap beans, *i.e.*, traces of both compounds after processing. FARROW *et al.* (1968) found that home washing and cooking removed 55 percent of the carbaryl residues in broccoli. LAMB *et al.* (1968 b) pointed out that after canning the DDT residues in spinach converted to other compounds but parathion and carbaryl residues did not change. Further home cooking had little or no effect on the residues.

YAO and GEISMAN (1972) demonstrated that boiling or cooking spinach in water for six to seven minutes caused complete decomposition of malathion residues. They also concluded that freezing caused no reduction of residues. LICHTENSTEIN *et al.* (1965) also indicated that boiling carrots for 30 minutes removed heptachlor but not aldrin. GEISMAN (1972) reported that dacthal residues in tomato juice were destroyed by flash pasteurization at 252°F for 0.7 second.

V. Discussion

Whether a pesticide residue could be removed or reduced by preparation and processing depends primarily on the fate of the pesticide and the unit operations employed.

a) Fate of pesticides

It is suggested that the fate of the pesticide depends on the method of application. If a material is applied to the soil, the main pathway of ingress to the plant would be through the roots with subsequent translocation, if possible, throughout the plant. If this were to occur, any unit operation which affects the exterior surface of the edible portion, e.g., washing or peeling, would probably cause little, if any, removal of the residue. Unfortunately, most pesticides that are applied to the soil are readily translocated throughout the plant. For those chemicals which are translocated, operations which subject the product to heat have potential for reducing residues.

If foliar application is utilized, another hypothesis must be considered. First, the edible portion would be a factor. With spray application, the highest residues would be found on the leaves. The concentrations found on fruits, while less, would depend on the amount of foliage cover and the nature of the fruit surface, *i.e.*, presence of a waxy coating or of pubescence. Stems would be similar to fruit but the underground portion would likely contain no residues.

Secondly, the persistence of the pesticide would affect the residue. Persistence is related to adsorption and absorption. A material that dissipates with a short half-life is usually adsorbed on the plant surface and is subject to rapid physical and chemical diminution by the environment. Those chemicals which are persistent exhibit this phenomenon but eventually are absorbed into the plant tissue where a second and longer half-life decrease occurs. This decrease is partially due to metabolism and partially due to dilution resulting from growth of the plant material.

A third consideration would be application rate. The relationship is a direct one. The rate of application would be important only when a chemical was misapplied, e.g., if applications overlapped.

The amount of pesticide residue on a food crop appears to be due to interactions of all the variables mentioned above. In general, the concentrations of nonpersistent pesticides applied to the crop should be related to the edible plant part and be a function of the length of time after application. On the other hand, concentrations of persistent pesticides should be a function of time assuming the proper rate of application in both cases.

b) *Removal by washing*

Washing appears to be the most important unit operation a processor can use to reduce or remove residues. Evidence has been presented that if the residue is adsorbed on the surface of the plant part then the probability for reducing the concentration by washing would be high. If the residue is absorbed into plant tissues, other unit operations might be utilized to remove it (GEISMAN and GOULD 1970).

It is significant to note that pesticide residues tend to harden or set-up with time. The practical implication is that if a pesticide is applied too near harvest or in too high a concentration, the processor has a means to remove it. The chemical can be largely removed if the commodity is washed within a short time after application, usually within one day.

c) *Removal by heating*

Most pesticides which appear to be heat stable may actually be degraded when heated in the presence of food products. Any unit operation which employs heat offers potential in reducing residues, if present. The processing flow sheets for many commodities include at least one such operation.

Home preparation and cooking of many products also aids in reducing or removing residues. The consumer can play a role in the safety of his own food supply in this manner.

VI. Conclusions

The unit operations employed in preparing and processing food play a role in reducing the amount of pesticide residue, if present. Washing has been investigated in many in-depth studies. This operation could be utilized on commodities for the fresh market. It seems that should large residues be present multiple washings could greatly reduce them. Other unit operations employing heat could be utilized to this end.

It also seems readily apparent that the effects of individual unit operations are additive. When a portion of a residue is removed by washing an additional portion is removed by blanching (FARROW 1969 b). Therefore, if a residue were detected then a series of washes might be given. Other operations could also be repeated.

An implication of most of the research reported herein is to investigate other unit operations for their potential in removing residues. Refinements of existing techniques and newly discovered methods should also be evaluated. The net result would be to assure the consumer of a safe wholesome food supply.

Summary

An adequate and safe food supply has highly emotional implications to society. Unit operations normally employed in processing food crops reduce or remove pesticide residues, if present. The washing operation has been studied in-depth with specific recommendations derived for many commodities. Other unit operations such as peeling, blanching, and cooking play a role in reducing residues. Each operation has a cumulative effect on the reduction of pesticides present. Refinements of existing operations and newly devised operations should be evaluated for efficiency in removing residues.

The consumer can contribute to the reduction of residues by home preparative methods. Most commodities can be washed at home and proper cooking destroys many residues. It would seem beneficial to include this information in consumer education programs.

References

BOHM, R. O., F. C. LAMB, L. D. LEWIS, and D. C. WHITE: Insecticide residue studies on raw and canned green beans. Nat. Canners Assoc. Res. Rpt. No. 13600-C (1950).

BRITTIN, W. A., and J. D. FAIRING: A study of the effects of processing on spray residues. J. Assoc. Offic. Anal. Chemists **33**, 599 (1950).

CARLIN, A. F., E. T. HIBBS, and P. A. DAHM: Insecticide residues and sensory evaluation of canned and frozen snap beans field-sprayed with Guthion and DDT. Food Technol. **20**, 80 (1966).

CARTER, R. H., P. E. HURANKS, H. D. MANN, L. M. ALEXANDER, and G. E.

SCHOPMEYER: Effect of cooking on the DDT content of beef. Science **107,** 347 (1948).

CHEN, S. C., J. L. COLLINS, I. E. MCCARTY, and M. R. JOHNSON: Blanching of white potatoes by microwave energy followed by boiling water. J. Food Sci. **36,** 742 (1971).

DURHAM, W. F.: Pesticide residues in foods in relation to human health. Residue Reviews **4,** 34 (1963).

ELKINS, E. R., F. C. LAMB, R. P. FARROW, R. W. COOK, M. KAWAI, and J. R. KIMBALL: Removal of DDT, malathion, and carbaryl from green beans by commercial and home preparative procedures. J. Agr. Food Chem. **16,** 962 (1968).

FARROW, R. P., E. R. ELKINS, and R. W. COOK: Conversion of DDT to DDE in canned spinach. J. Agr. Food Chem. **14,** 430 (1966).

——, F. C. LAMB, R. W. COOK, J. R. KIMBALL, and E. R. ELKINS: Removal of DDT, malathion, and carbaryl from tomatoes by commercial and home preparative methods. J. Agr. Food Chem. **16,** 272 (1968).

—— ——, E. R. ELKINS, R. W. COOK, M. KAWAI, and A. CORTES: Effect of commercial and home preparative procedures on parathion and carbaryl residues in broccoli. J. Agr. Food Chem. **17,** 75 (1969 a).

——, E. R. ELKINS, W. W. ROSE, F. C. LAMB, J. W. RALLS, and W. A. MERCER: Canning operations that reduce insecticide levels in prepared foods and in solid food wastes. Residue Reviews **29,** 73 (1969 b).

GEISMAN, J. R.: Reduction of pesticides from horticultural food crops by processing. Hort. Sci. **7,** 227 (1972).

——, and J. DEPPEN: Removal of DDT residues by unit operations in preparing and processing spinach. Ohio Agr. Res. Dev. Center Mimeo. Rpt. 337 (1967).

——, and W. A. GOULD: Physical and chemical factors affecting the efficiency of washing sweet corn. Ohio Agr. Expt. Station Res. Bull. 954 (1963).

—— —— Washing fruits and vegetables to remove residues. Ohio Agr. Res. Dev. Center Res. Circ. 173 (1970).

GOULD, W. A., J. R. GEISMAN, and J. P. SLEESMAN: A study of some of the physical and chemical factors affecting the efficiency of washing tomatoes. Ohio Agr. Expt. Sta. Res. Bull. 825 (1959).

HALLER, H. H., and R. H. CARTER: DDT spray residues on apples and the effect of removal treatments. Proc. Amer. Soc. Hort. Sci. **56,** 116 (1950).

KOIVISTOINEN, P., and A. KARINPAA: Stability of isopropyl N-phenylcarbamate (ICP) and isopropyl N-(3-chlorophenyl)-carbamate (CIPC) residues on fruit treated after harvest. J. Agr. Food Chem. **13,** 459 (1965).

—— ——, M. KONONEN, and P. ROINE: Magnitude and stability of captan residues in fresh and preserved plant products. J. Agr. Food Chem. **13,** 468 (1965).

KROCHTA, J. M., G. S. WILLIAMS, R. P. GRAHAM, and D. F. FARKAS: Reduced-water cleaning of tomatoes. Proc. Food Processing Waste Management Conf., Cornell Univ. (1973).

LAMB, F. C., L. D. LEWIS, and R. O. BOHM: Insecticide residue studies on raw and canned apricots. Nat. Canners Assoc. Res. Rpt. No. 13400-C (1950).

—— ——, and S. K. LEE: Studies on removal of insecticide residues from apricots. Nat. Canners Assoc. Res. Rpt. No. 12441-A (1948).

——, R. P. FARROW, E. R. ELKINS, R. W. COOK, and J. R. KIMBALL: Behavior of DDT in potatoes during commercial and home preparation. J. Agr. Food Chem. **16,** 272 (1968 a).

—— —— ——, J. R. KIMBALL, and R. W. COOK: Removal of DDT, parathion, and carbaryl from spinach by commercial and home preparative methods. J. Agr. Food Chem. **16,** 967 (1968 b).

LAZAR, M. E., D. B. LUND, and W. C. DIETRICH: A new concept in blanching (IQB). Food Technol. **25,** 684 (1971).

Lichenstein, E. P., G. R. Myrdal, and K. R. Scula: Absorption of insecticide residues from contaminated soils into five carrot varieties. J. Agr. Food Chem. **13**, 126 (1965).

Liska, B. J., and W. J. Stadelman: Effects of processing on pesticides in food Residue Reviews **29**, 61 (1969).

Mercer, W. A.: Radioactive tracer techniques for detecting and measuring detergent residues on washed foods and evaluation of food washing practices. Nat. Canners Assoc. Progress Rpt. D-949 (1960).

—— Application of radiochemistry techniques in food processing research. Nat. Canners Assoc. Bull. SAN-1020, Part 1 (1962).

Metcalf, R. L.: Methods of estimating effects. In C. O. Chichester (ed.): Research in pesticides, p. 17. New York: Academic Press (1965).

Ralls, J. W., H. J. Maagdenberg, N. L. Yacoub, and W. A. Mercer: Reduced waste generation by alternate vegetable blanching systems. Environ. Prot. Technol. Series EPA-R2-72-018 (1972).

——, W. A. Mercer, R. P. Graham, M. R. Hart, and H. J. Maagdenberg: Dry caustic peeling of tree fruit to reduce liquid waste volume and strength. Environ. Prot. Technol. Series EPA-R2-72-018 (1972).

Smith, E., A. L. Ryall, and C. R. Gross: The removal of lead, arsenic and fluorine residues from apples. Wash. State Hort. Assoc. Proc. **29**, 86 (1934).

Tressler, C. J.: The recovery of DDT from canned foods and its stability during processing. J. Assoc. Offic. Anal. Chemists **30**, 140 (1947).

Walker, K. C.: Problems relating to the removal of DDT spray residue from apples. J. Agr. Res. **78**, 383 (1949).

Wessel, J. R.: Pesticide residues in foods. N.Y. State Agr. Expt. Station Special Rpt. No. 9 (1972).

Yao, B. C., and J. R Geisman: Removal of malathion from spinach by commercial and home preparation procedures. Ohio Agr. Res. Dev. Center Res. Circ. 188 (1972).

Manuscript received November 26, 1973; accepted January 21, 1974.

Consequences of lead in the ambient environment: An analysis

By

Robert M. Bethea and Nancy J. Bethea*

Contents

I. Introduction

This report is composed of two parts: a discussion of the real and potential problems associated with airborne lead and an analysis of the position of the *U.S. Environmental Protection Agency* (EPA) in support of a national ambient air quality standard of two μg of

* Chemical Engineering Department and Department of Biological Sciences, respectively, Texas Tech University, Lubbock, Texas 79409.

lead/m³. EPA proposes to achieve this standard by controlling the use of lead additives in gasoline.

II. The problem of airborne lead

a) Sources of atmospheric lead

The sources of atmospheric lead are the manufacture and use of pesticides; refuse incineration; coal combustion; production of chemicals, caulking compounds, paint pigments, solder, cable covering, ammunition, and storage batteries; manufacture of glass and ceramics; and combustion of fuels containing lead additives. All of these are stationary sources except the last. Precise lead emission rates are not readily available for any of the fixed sources. Lead emissions from mobile sources have been estimated as being in excess of 180,000 tons/year (BAZELL 1971). An estimate (MCCAULDIN 1966) of lead emissions for the Cincinnati area showed that the amount of airborne lead from gasoline consumption was approximately 30 times greater than that from coal combustion. Given the average concentrations of lead in the two classes of fuels, it is reasonable to expect a similar emission ratio in all large metropolitan areas using fossil fuel for power generation. When the average lead content of coal [ten ppm as determined by CHOW and EARL (1970 a)] and that in leaded gasolines (two to four g/gal) were compared in view of the relative consumption of these fuels, a similar conclusion was reached: that the contribution of lead from coal combustion is negligible when compared to the amounts of airborne lead in the San Diego area produced from leaded gasoline (CHOW and EARL 1970 b). Although specific, corroborating data were not found for other cities, there is no reason to doubt the applicability of these conclusions to other, similar U.S. metropolitan areas with similar meteorologic and climatic factors.

b) Lead concentrations in the atmosphere

It has been postulated (GOLDWATER and HOOVER 1967) that atmospheric lead emissions from insecticides, lead pipe, and paint have been decreasing since 1960. The average lead content of urban atmospheres has remained relatively constant according to data collected by the National Air Surveillance Network (*U.S. Department of Health, Education, and Welfare* 1966 and 1968; *U.S. Environmental Protection Agency* 1971 and 1972). The average annual lead concentrations in the air were one to three $\mu g/m^3$ for most cities, 0.1 to 0.5 $\mu g/m^3$ for nonurban locations, and less than 0.1 $\mu g/m^3$ at remote sites. In one study (LANDAU *et al.* 1969) about 50 percent of the variability of atmospheric lead was attributable to fluctuations in gasoline sales; the remaining variation was thought due to changes in traffic density and wind near the sampling sites.

c) *Exhaust lead characteristics*

One estimate of vehicular lead emissions (HIRSCHLER and GILBERT 1964) was that 20 to 55 percent of the lead consumed by the engine is exhausted and that the major factors in actual emission levels are speed, prior driving history, and exhaust system condition. These factors have been confirmed and the range of the fraction of intake lead widened as a result of more recent work. Approximately 20 to 75 percent of the lead additives in gasoline is exhausted, depending on driving conditions (HALL 1972). Another estimate (TER HAAR *et al.* 1972) is that only seven to 30 percent of intake lead is emitted and that the actual amount varies with the condition of the exhaust system.

Lead is normally emitted from gasoline combustion in the form of lead halides, chiefly as lead chlorobromide or as a complex of lead halides with ammonium halides (HIRSCHLER and GILBERT 1964, TER HAAR and BAYARD 1971, HALL 1972). The halides come from the exhaust system scavengers, ethylene dichloride and ethylene dibromide, added in the gasoline. After 18 hours, under laboratory conditions, about 75 percent of the bromides and about 30 to 40 percent of the chlorides appear to be converted to lead carbonates, lead oxycarbonates, and lead oxides, all of which are virtually insoluble in water at ambient temperatures. Small amounts of lead sulfate and lead phosphate can be found as a result of reaction with sulfur trioxide in the atmosphere and any phosphorous in the fuel. These experimental results (TER HAAR and BAYARD 1971) have striking similarities to atmospheric samples. The conclusion from these reports is that significant amounts of inorganic lead are emitted into the atmosphere from combustion of motor fuels containing lead additives and that this lead is rapidly converted under basic conditions to insoluble forms.

d) *Determination of sources of airborne lead*

An important investigation (CHOW and EARL 1972) has recently shown that lead can be differentiated as to geologic source by examination of the isotopic ratios, $^{206}Pb/^{204}Pb$ and $^{206}Pb/^{207}Pb$. The value of these isotopic ratios for any lead sample can distinguish it as originating from ore, coal, or sedimentary deposits. A previous study had already shown that, within the limits of experimental error, atmospheric lead could be (and was) related to the leaded gasoline being sold in several world-wide metropolitan areas (CHOW and JOHNSTONE 1965). This technique was then used (CHOW and EARL 1970 b) in a 39-month study in San Diego which showed a definite relationship between isotopic ratios of aerosol lead and local gasoline lead. This study disclosed two other pertinent facts: the aerosol lead content in the San Diego atmosphere was increasing by five percent per year and that seasonal maxima and minima in airborne lead occurred in January and July, respectively.

e) Deposition of airborne lead

The deposition of airborne lead on the earth's surface and its incorporation into the human food chain is not open to question. The concentration of lead in the earth's crust has been estimated as ten to 15 ppm (CHOW and PATTERSON 1962) and 10 ppm (MCCAULDIN 1966, DE TREVILLE 1964). Deposition of aerosol lead is presumably by rainout and washout. The average lead concentration in precipitation samples obtained through the National Air Surveillance Network was 36 μg/l. (LAZRUS *et al.* 1970). In La Jolla, California, average lead in precipitation has been measured (CHOW and EARL 1970 b) at 40 μg/l. The investigators' observation that there was an inverse relation between amount of rainfall and its lead concentration is worthy of note.

A comparison of La Jolla and San Diego dusts indicates that only a fraction of the fine lead aerosol precipitates near the source; the remainder is dispersed world-wide by the winds (MUROZUMI *et al.* 1969).

Evidence of accumulation of lead on the earth's surface has been provided (MUROZUMI *et al.* 1969) by studies of lead deposited in Greenland snow layers. A drastic, approximately three-fold increase was found for deposits laid down between 1940 and 1950. This lead could only have come from airborne sources whose emission points are hundreds of miles distant. Isotopic ratio examination showed (TATSUMOTO and PATTERSON 1963 b) that lead accumulated in rural snow in the Lassen Volcanic National Park during the winter of 1962 is similar to that of the Los Angeles basin aerosol.

The probable significance of these remote lead deposits is disturbing in that the investigators have presented two convincing arguments that 99 percent of the lead aerosol produced at temperate latitudes in the northern hemisphere should have been removed by washout before reaching these remote regions. The implication of their reasoning is that the surface lead concentration over the northern hemisphere is increasing at an unacceptable rate.

Lead concentrations in surface water samples of the Pacific Ocean (TATSUMOTO and PATTERSON 1963 a) and of the Atlantic Ocean and Mediterranean Sea (TATSUMOTO and PATTERSON 1963 b) have been measured. It was postulated by the investigators that the bulk of the lead found originated from wind-dispersed lead aerosol from automotive emissions, there being for the most part no natural or artifactual emissions within several hundred miles. As the lead concentration in snow samples reported above and in rainfall (LAZRUS *et al.* 1970) in the U.S. is about ten to 100 times greater than found in the surface samples on the open seas, the fraction of airborne lead rapidly removed by washout appears confirmed.

f) Lead accumulation in water supplies

In 1964 the average concentration of lead in public water supplies in this country was 0.01 ppm (DE TREVILLE 1964). The standard for lead in potable water is 0.05 mg/l. which is approximately 0.05 ppm. In a long-term study (ETTINGER 1966), only five out of 876 municipal water samples collected in a nationwide survey exceeded the potable water lead standard. Some of the stations whose samples did not show appreciable amounts of lead were the Delaware and Schuylkill Rivers at Philadelphia, the Hudson River below Poughkeepsie, the Connecticut River at Enfield Dam, and the Illinois River at Peoria. To quote ETTINGER, "at these sites the combustion of coal, the consumption of gasoline, or massive aggregations of plumbing and industry, have had maximum opportunity to enrich the lead content of streams, but lead was not found in large amounts in solution." From this report and the fact that, as discussed below, lead once deposited on surface soil is relatively immobile, it is concluded that airborne lead has not yet affected the lead content of municipal water supplies. It should be pointed out that no bottom samples were taken for lead assay and that the potable water standards are concerned only with soluble lead, not total lead.

g) Lead accumulation in roadside soil and vegetation

Many studies have been conducted to ascertain the degree of lead buildup in roadside soil and vegetation. In the earliest work found, vegetables grown within 25 ft of heavily travelled roads had lead concentrations averaging 80 to 150 ppm (CANNON and BOWLES 1962). These investigators commented that the amount of lead in the plants was dependent on traffic volume and distance from the roadway. Another study (SINGER and HANSON 1969) of 12 highways in the Minneapolis-St. Paul area showed that roadside soil and vegetation at sites near roads with equivalent traffic density were similar in lead content. These data showed a strong indication that lead in roadside soil decreases with distance from the roadway and soil depth, and that the concentration is dependent on the time the roadway has been in use. A different study (SCHROEDER and TIPTON 1968) showed that lead remains in the soil a long time after deposition and is only slowly eluted by rain. Other studies have confirmed the decrease in lead concentration with distance from the road and with depth in the soil profile. These studies were not designed to show any dependency between lead concentration in soil or roadside grasses and traffic density (LAGERWERFF and SPECHT 1970, TER HAAR and BAYARD 1971, LEE 1972). Another study (CHOW 1970) also showed a strong correlation between lead concentration and soil depth and distance from the road. This study showed that lead content of roadside grasses was two to 12 times the concentration of the soil in which they were growing. Isotopic ratio examination of these grass and soil samples have provided positive proof that the excess lead in surface soil originates in gasoline exhaust and that the lead found in the grass is similar to that in the surface soil. From all of the above data, it is concluded that lead deposits are relatively immobile, that they apparently are not greatly affected by precipitation or runoff, and that lead is definitely accumulating in surface soil layers adjacent to heavily travelled roads.

h) *Lead in urban soils*

The number of studies on the lead concentration in urban foliage and soil is limited. A report (BAZELL 1971) on lead poisoning in animals in the Staten Island (New York) Zoo showed that lead found in the soil and ground cover of the zoo equalled or exceded that found along major highways. Another urban study (SMITH 1972) presented data for the lead content of woody plants growing within eight meters of streets. These concentrations were more than 100 times the levels found in similar plant tissue in the natural, unpolluted environment. The soil in the general vicinity of these trees contained 35 to 50 times the average amount of lead found in the earth's crust. These data and that of the last roadside study (CHOW 1970) mentioned above are clear evidence of the concentration of airborne lead in plants. This lead may be returned to the soil and the atmosphere in a relatively short time by the mechanism of decay or burning of dropped leaves, twigs, etc. (SMITH 1972).

There is thus no doubt that lead can accumulate in the leafy parts of large as well as small plants. The concentration of lead in the leaves is particularly disturbing as it forebodes the accumulation of this poison in forage crops unless emissions are drastically reduced.

i) *Lead in human food*

Most of the data available on the lead content of human foods was obtained prior to 1940 with sporadic data from 1960. Lead contents of various meats, milk, poultry, vegetables, and fruits were summarized in an earlier review (DE TREVILLE 1964). The lead contents in meat, milk, and poultry were generally below 0.3 ppm with some notable exceptions. These were beef liver 0.42 ppm, beef kidney 0.35 ppm, ground beef 0.6 ppm, pork sausage up to 1.6 ppm, fish up to 0.54 ppm, and shellfish and crustaceans up to 3.0 ppm. It was noted than animal flesh tends to have significantly higher lead concentrations than do foods of vegetable origin. The results of KEHOE *et al.* (1940) suggest that the lead content of cow's milk is a function of the total body lead burden rather than of the daily diet. If true, milk lead should serve as an integrated measure of the exposure of cows to lead. Data obtained prior to 1942 showed the lead content of milk ranged from 0.02 to 0.14 ppm. The source of the lead was felt to be bulk milk cans in common use at that time. More recent analyses of milk processed in a modern dairy showed 0.02 to 0.07 ppm of lead.

Our knowledge of lead in vegetable and fruit crops grown for human consumption is predominantly from 1940. The summary available (DE TREVILLE 1964) shows that the concentrations of lead in most vegetables are below 0.1 ppm. Some notable exceptions were seedling maize 0.3 to 18 ppm, cabbage 0.5 to 3 ppm, cauliflower two ppm, beets 0.2 to six ppm, and lettuce one to nine ppm. Dried fruits and vegetables were unfortunately lumped together and reported as containing one to two ppm. The lead content of herbs was considerably higher than for vegetables in general. This, plus the higher concentrations in leafy vegetables and the concentrating action of roadside grasses, suggests that lead may be preferentially concentrated in the leafy parts of the plant rather than in fruit, stem, and root. Since leafy vegetables make up a small portion of our total diet, but we feed our cattle large amounts of foliage, we might expect meat and milk to become a much larger source of lead in the near future.

j) Lead poisoning in plants and animals

As a result of their study on corn mitochondria (KOEPPE and MILLER 1970), it was speculated that the effects of lead on plant growth may be accentuated if the plant is deficient in phosphate. Although no evidence has been reported regarding the effects of lead exposure for domestic animals suffering from any dietary deficiency, it is reasonable to suppose that this phenomenon would be revealed by controlled experiments designed to elucidate such effects.

An interesting effect of lead poisoning in urban pets has recently been reported (ZOOK *et al.* 1972). The high frequency of plumbism in urban dogs less than one year old was not shared by cats. The investigators assumed these results to be evidence that airborne lead is not likely to be the sole source of poison for animals and man. They noted that the effects of blood lead on δ-aminolevulinic acid dehydrase function were similar in dogs to the effects reported for human beings. Their conclusions may be at fault. If the old night-time adage, "bring in the dog, put out the cat," has any validity, then dogs are exposed to the outdoor environment during daylight hours and cats are predominately exposed, if at all, to the outdoor environment during the hours of darkness. Because of differences in urban traffic patterns with time of day, the exposures of these two species of household pets would then be vastly different.

One significant incidence of lead poisoning in large animals has been reported (SCHMITT *et al.* 1971). Horses in the vicinity of the large lead smelter near Trail, B.C., suffered lead poisoning due to high lead levels in their forage. Lead content of this forage has been related to accumulated lead in surface soil from nearby smelter emissions. Contemporary analysis of the ambient air, water, and locally grown human foodstuffs showed no evidence of any significant human health hazard when the lead contents of those samples were compared to the corresponding Canadian standards for 1969.

k) Lead poisoning in man

Lead poisoning in human beings occurs in two primary ways depending on the type of lead compound involved. Lead intoxication, which is chronic poisoning, occurs predominantly due to the absorption of lead oxide, lead carbonate, and other water-insoluble lead compounds in the gastrointestinal tract. Acute inorganic lead poisoning is exceedingly rare and results almost exclusively from the inhalation of fine particles of lead oxide (FELTON *et al.* 1972). Organic lead poisoning is usually caused by inhalation of tetraethyl lead which is both highly volatile and lipid-soluble. The amount directly absorbed through the skin is probably insignificant compared to respiratory absorption because of the differences in the surface areas involved.

l) Alimentary absorption of lead

Having shown how lead emissions may be dispersed into and deposited from the atmosphere it is pertinent to consider the mechanism by which lead is ingested into and absorbed by the human body. The average daily lead intake from all sources has been estimated at 300 μg/day.

The result of one study (GOLDSMITH 1969) indicated that approximately equal amounts of lead are absorbed into the body by the alimentary and respiratory routes. Isotope studies showed (MORSE and WELFORD 1971) that the dietary intake of ^{210}Pb in New York City is higher than the intake by inhalation. However, dietary lead is less easily assimilated than airborne lead. Thus the result is that the blood lead level derived from these sources is approximately equal. This work corroborates earlier studies which showed that less than ten percent (30 μg/day) of ingested lead is absorbed in the gastrointestinal tract (KEHOE 1961 and 1968, *Anonymous* 1971, GOLDSMITH and HEXTER 1967 and 1968). Other estimates of the fraction of ingested lead absorbed through the gut are eight percent (MORSE and WELFORD 1971) and six-to-seven percent (HALL 1972).

KEHOE (1969) showed that the larger factor causing elevated blood lead in the general population under ambient exposure is alimentary absorption. This has been corroborated by GRUNDY (1969) who showed that 66 percent of the total body lead burden is due to ingestion. A prior review of the biological effects of lead (HARDY 1969) disclosed that at that time no relation between the alimentary absorption rate and subject age had been reported.

m) Absorption of respiratory lead

Inhaled lead particles are deposited by three primary mechanisms: inertial impaction, sedimentation, and Brownian motion. The amount and site of deposition depend on particle size, density, hygroscopic nature, and such physiologic functions as ventilation rate and depth and duration of exposure (TOMASHEFSKY and MITCHELL 1966). The absorption rate is determined by the nature of the material and its solubility in water or lipids. Approximately 50 percent of inspired lead is retained in the lungs if the particle diameter is at or below one μ (HALL 1972). The remainder of these small particles is expired. Particles greater than two μ in diameter impinge on the mucus membranes of the upper respiratory tract and are cleared by ciliary action and swallowed or expectorated. If swallowed, they behave as if originally ingested in the diet (HALL 1972). No demonstrable portion of airborne lead is deposited in the upper respiratory tract and then diverted into the gastrointestinal tract when fine (90 percent below 0.17 μ) particles are inhaled. For larger particles (larger than one μ) transfer of lead from the respiratory system to the gastrointestinal tract has been well-established (KEHOE 1964) by comparing dietary lead with fecal lead.

n) Significance of blood and urinary lead levels

The concentration of lead in blood is not necessarily well correlated with biochemical evidence of metabolic effects (GOLDSMITH and HEXTER 1967). Concentration increased due to alimentary and respiratory absorption of lead and its compounds and varied directly with exposure and dose (KEHOE 1968) under controlled experimental conditions. The "normal" concentration of lead in blood ranged from 15 to 40 μg/100 ml of whole blood with an average of 17 μg/100 ml of whole blood (GOLDWATER and HOOVER 1967). Blood lead concentrations greater than 40 μg/100 ml of whole blood were considered (STEINFELD 1970) evidence of undue lead absorption. Epidemiologic data showed that there has been no significant change in adult blood lead levels in the U.S. between 1925 and 1965 (STOPPS 1965).

A comparative study of primitive peoples in five remote locations and 6,000 American men and women (with yearly sampling in 25 locations) showed no difference in blood lead level between remote and urban populations (STOPPS 1969). This is in direct conflict with previous comparisons between urban adults and children *versus* their rural counterparts in the U.S. Those surveys showed slightly higher urban adult blood levels (GOLDWATER and HOOVER 1967) while urban children showed blood lead levels almost twice that of rural children (HAGGERTY and NORLAND 1969).

The normal range of urinary lead was 20 to 60 μg/l. with a mean value of 35 μg/l. (GOLDWATER and HOOVER 1967). Examination of urine samples in conjunction with a 40-year study of blood lead levels referred to above (STOPPS 1965) showed no significant change in the normal value of urinary lead during that period. A few human experiments have been conducted to determine the effects of exposure to simulated occupational lead levels. The most significant characteristics of these deliberate intermittent respiratory exposures was an increase in urinary lead excretions over a period of several months to a maximum value which was not subsequently exceeded (KEHOE 1964). A controlled experiment simulating both ambient and occupational exposures (KEHOE 1968) showed that the urinary lead excretion rate was directly dependent on exposure and dose. For deliberate exposures to occupational levels of airborne lead the average concentration was well correlated with severity of exposure (STANKOVIĆ 1971).

o) *Biochemical effects of absorbed lead*

The best-documented example of lead inhibiting enzyme reactions in man is δ-aminolevulinic acid dehydrase (ALAD). The inhibition of this enzyme is physiologically significant in adults when blood lead levels are 40 μg/100 g and above. The degree of inhibition of this enzyme is usually related to the presence of δ-aminolevulinic acid (ALA) excreted in the urine (KING 1971, ZIELHUIS 1971). The concentration of ALA in urine for normal adults in the general environment was less than 4.5 μg/l. (STANKOVIĆ 1971). It should be pointed out that "normal" refers to the amount of any material found in persons who have had no evident occupational, medicinal, or unusual sources of exposure to lead (GOLDWATER and HOOVER 1967). The general environment refers to exposures to ambient rather than occupational levels of lead.

Probably the most succinct explanation (CHISHOLM 1971) of the effects of lead in the blood of human beings is as follows:

"Arithmetic increases in blood lead concentration above 40 μg.Pb/100 g. whole blood are correlated statistically with exponential increases in the daily urinary output of δ-aminolevulinic acid. There is a wide body of experimental and clinical data which indicates that the increase in ALA in urine is correlated with the inhibition of the activity of δ-aminolevulinic acid dehydrase in the tissues. Consequently the increase in urinary excretion of ALA provides evidence of an adverse metabolic effect; namely, interference in porphyrin metabolism and in the formation of heme. Heme is the prosthetic group of hemoglobin and the respiratory enzymes; i.e. the cytochromes found in all tissues. This adverse metabolic affect is observed to increase as blood lead concentration rises above 40 μg.Pb/100 g. whole blood."

Further discussions of the effects of lead on the human body may be found in several recent review articles (ZIELHUIS 1971, DE BRUIN 1971, HARDY *et al.* 1971, REDDICK 1971, ALBAHARY 1972). These reviews provide an enormous amount of detail regarding the metabolic effects of lead and its compounds on the human body. They discuss and amplify the rationale for the choice of measurement of lead in blood and urine and of ALA in urine as valid estimates of the amount of lead in the body and its chronic or acute origin. A considerable number of case studies are presented to illustrate their major points. They show that the common ultimate effect of excessive lead in the human body is depression of the central nervous system. It cannot be said, however, from the studies discussed, that lead is one of the causative agents of mental illness. The data in this regard are too scanty and lack adequate controls, as usually recognized by the original authors.

p) Discussion

From the data presented in this analysis of the environmental and particularly the human health aspects of lead, it is apparent that the continued uncontrolled emission of lead into the ambient air constitutes a serious problem. It is felt that conclusive evidence has been presented to show the accumulation of lead in leafy vegetables, grasses, and forage crops. Whether this accumulation is due to absorption from the soil or has been caused as a result of atmospheric fallout is immaterial. Complete elimination of lead antiknock compounds used in gasoline will not materially affect the lead content of foods in the immediate future. Such removal of one source of airborne lead is highly recommended. If airborne lead is significantly reduced in urban areas by the promulgation and enforcement of the proposed two $\mu g/m^3$ standard, then a considerable health risk should be decreased for a large portion of the population. Although the emissions of tetraethyl lead and tetramethyl lead appear to consititute the major source of airborne lead, any attempt to achieve the ambient air concentration proposed by EPA must, to be more immediately effective and to avoid undue economic stress to the petroleum and automotive industries, also include a corresponding reduction in lead emissions from all stationary sources.

Unless immediate and effective action leading to the control of atmospheric lead from all sources is initiated now, we run the definite risk of contaminating our agricultural lands and waters to the same degree (and with the same result) as soil and vegetation adjacent to heavily travelled roads in the forseeable future. To sum up, there is already too much lead in our environment; allowance of continuing emissions at present levels is foolhardy.

III. Conclusions I: Scope and nature of the airborne lead problem

Based on the foregoing discussion of the sources, degree of dispersion, and health-related effects of airborne lead emissions, the following conclusions seem inescapable: (1) the concentration of lead in the atmosphere from all sources is constantly increasing; (2) the majority of airborne lead is removed by washout and rainout over the major continental land masses; (3) lead deposited on the earth's surface, whether by sedimentation or precipitation, is relatively immobile, and results in an increasing concentration in surface soils; (4) grasses and other plants concentrate lead in their tissue by a factor of two to 12 compared to the soil in which grown; (5) domestic animals can, under conditions of unusual lead content in forage, exhibit all the classical problems of lead poisoning; (6) the mechanism of the biochemical involvement of lead in the human body is through interference with the formation and degradation of hemoglobin and erythrocytes, and by depression of the central nervous system; (7) the best measure of short term lead exposures is the lead content in blood; (8) urban populations appear to have higher blood lead levels than suburban or rural populations; (9) no epidemiological data are available showing a clear cut and unambiguous relation between elevated blood lead levels, ambient air lead concentrations, and frequency of lead-related disorders or diseases; and (10) further accumulation of lead in the ambient environment is intolerable.

Lead emissions from all sources should be uniformly reduced by whatever means necessary commensurate with the viability of this country's economy.

IV. Analysis of EPA's position on the health effects of airborne lead

In the EPA position paper on the health effects of airborne lead (*U.S. Environmental Protection Agency* 1972 b), the possibility is raised (Section II, p. 1) that lead intoxication or lead poisoning may cause mental retardation. The source of lead in both studies referred to by EPA was attributed to ingestion of lead-based paint. Another study referred to by EPA (Section II, p. 3), for which the source(s) of lead were not given, raised the possibility that minimal brain damage in the form of hyperactivity could occur. Even if the speculation "that children may be suffering subtle but unrecognized neurological impairments due to lead" is true, no evidence was presented that the lead involved originated from automotive emissions.

The problem of children with pica ingesting dirt and street dust containing high amounts of lead was brought out in the EPA position paper (Section IV, p. 2). Some of that lead, especially lead in yard dirt, undoubtedly originated from the deterioration of lead-based house paints. Some could as easily have come from fallout from automotive exhausts or from lead-using industrial operations. The EPA position paper virtually insists (Section V, pp. 1, 2, 10) that lead fallout from automotive exhausts is the major source of lead in urban street dust, parks, and yards. This EPA conclusion is supported by the lead washout rate analysis (MUROZUMI *et al.* 1969) previously discussed and by the examination of the lead content of urban woody plants, soil, and vegetation (SMITH 1972, BAZELL 1971). It is also reasonable to suppose that lead accumulates in city street dust and in soil adjacent to the streets in the same manner in which it builds up in roadside soils and grasses (CANNON and BOWLES 1962, SINGER and HANSON 1969, CHOW 1970, LAGERWERFF and SPECHT 1970, TER HAAR and BAYARD 1971, LEE 1972).

EPA states (Section V, p. 10) that lead concentrations up to 0.5 percent (5,000 ppm) are common in the fine fractions of street dust collected from busy roadways. The implication is that this lead comes from automotive exhaust. They also cited window-sill dust lead levels of 0.175 to 0.2 percent (1,750 to 2,000 ppm) with the same implication. That this lead exists is not questioned. What is questioned are the high concentrations involved and the implication of automotive exhaust as the prime (if not the sole) offender. The roadside soil studies referenced immediately above showed maximum lead content of 80 ppm at 75 ft away, 700 ppm at 4.5 ft away, 122 ppm at 25 ft away, and 68 ppm at 26.2 ft away from the road. The conclusion seems clear: factors other than vehicular emissions are responsible for high lead contents in street dust. This is all the more plausible in that lead in roadside dirt is relatively immobile and street dust can be, and usually is, washed away periodically by street cleaning and storm runoff.

It would appear logical to perform an isotope ratio study (CHOW and JOHNSTONE 1965; CHOW and EARL 1970 a and b, 1972) to determine the source of the lead in street dust and yard dirt in urban areas where there is a high incidence of lead intoxication or poisoning in children.

The position of EPA regarding the actual human health effects of lead (Section III, pp. 1, 2) at low levels of exposure would be greatly enhanced if data were available showing that the action of δ-aminolevulinic acid dehydrase (ALAD) in hemoglobin synthesis is actually inhibited in man. Testimony cited by EPA (KAHN 1972) indicated that the required evidence had been obtained by MAJIC but had not yet been published. Diligent scrutiny of *Cumulated Index Medicus* through August, 1973 did not disclose this publication regarding *in vivo* demonstration of ALAD inhibition. If the data were available from one of Dr. KAHN's associates, one wonders why EPA has not obtained and shown this crucial evidence. The inhibition has been well demonstrated *in vitro*. If the effect is the same in man, then it is reasonable to agree with the supposition that lead-induced inhibition of ALAD will grossly affect cytochrome synthesis.

Of critical importance to the EPA position is whether or not any of these effects are (or can be) associated with exposure to lead at low levels (*ca.* two $\mu g/m^3$ or less). No reports were found relating lead poisoning and/or blood or urine lead composition or urinary δ-aminolevulinic acid (ALA, indicative of ALAD inhibition) to exposure to the general atmospheric environment. Of key importance to EPA's position (Section V, pp. 2–8) is the demonstration of an actual relationship corresponding to lead concentration in the blood. Measurement of these two variables should be taken at the same time, and *not* eight to nine years apart as done in one study (HUNT *et al.* 1970) referenced by EPA (Section V, p. 2).

The use of blood lead as the primary exposure index by EPA (Section IV, p. 1) is reasonable. It would be appropriate, however, to point out that the difference in effect of lead stored in the body may be quite different from that due to recent or current exposure. If stored lead can be mobilized and transferred from the bones into soft tissues as a result of physiologic stress, then blood lead becomes a less reliable index of current exposure effects. Extreme care should be taken to exclude from surveys and experiments any persons recently stressed or likely to be stressed during experimental exposure in any way which would cause mobilization of stored lead if the most reliable results are to be obtained.

The weight of evidence supports the EPA blood lead level guidelines (Section IV, pp. 5, 6) except that no epidemiologic data have been presented which relate blood lead in babies and young children to atmospheric exposures.

With regard to the "Seven City Lead Study" (TEPPER and LEVIN 1972), the EPA position paper (Section V, pp. 2, 3) states that the method of data analysis governs the existence of a blood-lead/air-lead relationship. This is nonsense. The statistical procedures to be used in interpreting the result of any experiment or survey are governed by the corresponding experimental or survey design employed. Changing the method of data handling or analysis after the fact can lead to gross inconsistencies. EPA goes on to say that lead in local water supplies (and presumably in diets) may be significant and admits that "one can never be completely certain that increased blood lead levels in female urban residents are entirely due to increased air lead exposures."

Regardless of the controversy surrounding the results of this study, a significant elevation in blood lead of urban over suburban or rural populations is cause for concern. Other studies (STOPPS 1965 and 1969) presented data to the effect that world-wide blood and urine lead levels are approximately constant and have been so for many years. If true, this would lend credence to the cardiovascular studies (HUNT *et al.* 1970) previously mentioned in which the air and blood lead levels were determined eight to nine years apart. However, if blood and urine lead levels remained constant between 1925 to 1965, a period which spans 40 of the 50 years that lead anti-knock agents have been used in gasolines, what basis does EPA have in asserting that these fuel additives cause elevations in blood lead? If these reported results are true, then lead emissions from stationary sources must have been decreasing at the same rate that mobile source emissions have been rising.

The speculation raised by these data collections can only be resolved by a nationwide lead emissions inventory based on actual emission sampling, not the use of emission factors, and by a corresponding epidemiologic study. The use of emission factors is dangerous. They are properly valid only for the plants and process conditions under which they were obtained. Furthermore, the data available (*U.S. Environmental Protection Agency* 1972 c) are incomplete: no emissions were shown for glass and ceramic manufacture, bearing wear, agricultural chemicals, ammunition manufacture, re-emission from burning of city leaf and yard trash, production of caulking compounds, use in plumbing, etc. It should be further noted that the lead emissions listed were all estimated from emission factors. One such study (AZAR *et al.* 1972) has already been performed. In the original paper, the authors state, "adjustments failed to explain the differences (in blood and urine response, R.M.B.), suggesting that variables other than air lead, smoking habits and osmolality are affecting the data. One obvious missing variable is ingested lead". Careful scrutiny of this preprint did not disclose any attempt by these investigators to measure the lead intake by food and water for their subjects. Yet the EPA position paper (Section V, pp. 6, 7) states about this work that "when all groups were combined and adjusted statistically to account for dietary lead differences among groups, a significant correlation between air lead and blood lead was obtained". It appears that EPA has adjusted the results to suit its own ends by the use of nonexistent data. This is such a blatant violation of the rules of scientific inquiry that its very occurrence subjects the entire EPA position paper to grave doubts regarding its validity. That these data were warped is bad enough. That this action was not caught in the EPA internal review process is unforgivable.

Furthermore, the authors (AZAR *et al.* 1972) stated clearly that their data showed a lack of any significant correlation between air lead and blood lead and advanced four reasons for this. These were: nonexistance of such a correlation; other, undetermined lead sources;

data range too small to show any significant effect if it does exist; and a sample size which was too small. EPA chose to insist that the sample size, 30 subjects per group, was too small, and that the study was therefore of little value. This is difficult to accept in that EPA "adjusted" the data as described above so as to pool it for use.

The authors further showed by analysis of the slopes of the regression lines, presumably by Bartlett's test for variance homogeneity or by appropriate F-tests, both of which are standard methods for this type of comparison (OSTLE 1963), that the data on all groups could be pooled directly. No "adjustments" were made by the authors. The result of this pooling showed the slope of the pooled regression line to be significantly different from zero but that less than 50 percent of the variability in blood lead was *not* accounted for by air lead variation. Common engineering practice demands a correlation of 90 percent or better for acceptability of the existance of a relationship between variables. As a result of the low correlation obtained, the authors stated that use of their regression line is improper because lead intake by sources other than air were not included in the study or data analysis.

For this study, all of the subjects were exposed to air lead concentrations as high as 9.12 $\mu g/m^3$ and all but one had blood lead levels below 40 $\mu g/100$ g of whole blood. The high blood lead in the other subject was attributed to other factors. This study is the only one reported to date in which the subjects wore personal air lead monitors and had blood lead determined at regular intervals. As such, its validity is considerably higher than the "Seven City Study" or the "Roadway Study" (DAINES *et al.* 1972). Relegation of this study (AZAR *et al.* 1972) by EPA to the status of an immaterial effort is unacceptable.

EPA has taken other data out of context. The "Roadway Study" cited above was conducted in two parts. In the first, women living 33.5 and 475 m away from a major traffic artery (turnpike with 60,000 vehicles/day) showed no significant differences in blood lead. The second part of this study (subjects living near a one-way street with 30,000 vehicles/day) showed that the blood lead was higher for subjects living 38.1 m away than for subjects living 121.9 m away. All subjects had lived in their homes at least one year. The study showed the concentrations of airborne lead to be greatest nearest the highway (3.7 m away) and significantly reduced at 38.1 m. Beyond that point, there was no significant difference in air lead concentrations between 38.1 and 121.9 m away from the road. By extracting the data for women having blood lead greater than 29 $\mu g/100$ g of whole blood, EPA gives the impression (Section V, p. 4) that elevated blood lead may be expected in women living nearer heavily travelled roadways than in women living at more distant locations. It is doubtful that the authors of this study (DAINES *et al.* 1972) intended that such inferences be made.

The results of the "Prisoner Study" (KNELSON *et al.* 1972) were used by EPA (Section V, p. 8) in an attempt to show that continuous exposure for 12 weeks to air lead concentrations of about 3 $\mu g/m^3$ can cause a 30 percent elevation in blood lead. This entire study and all its results are invalid for the following reasons. Blood volume was not measured but was assumed constant at 4.5 1. for each subject. A fixed ventilation rate of 15 m^3/day was assumed. This corresponds to a resting condition, rather than light exercise (GUYTON 1966, SCHMIDT-NIELSEN 1970) or even the prison environment. Lead absorption was fixed at 37 percent of that inspired for all subjects. Certainly these parameters could have been estimated more closely for each individual subject. A more serious defect was the changing subject population. No indication was given that the data were adjusted for this. In addition, the implication was given that blood lead was determined twice only for each subject: before and after the exposure. Use of the data from this study is clearly not warranted.

Another study of lead in children thought not to be exposed to leaded paint (MCINTYRE and ANGLE 1972) showed elevated blood levels between children in "high" (1.69 $\mu g/m^3$) vs. "low" (1.48 $\mu g/m^3$) average lead atmospheres. The distinction between "high" and "low" atmospheric lead concentration appears to have been made for the purposes of explaining the results. The authors pointed out that the differences in airborne lead may not have been significant but that dustfall lead may have been the dominant factor. The amounts of lead emissions from the several nearby industrial sources were not measured. No determination of the relative contributions to airborne lead level from those sources as opposed to automotive sources was made. The results appear to have questionable validity as used by EPA (Section V, pp. 8, 9).

The EPA position paper (Section V, p. 12) shows that calcium dietary deficiency may promote absorption of atmospheric lead in the rat. As other work (ZOOK *et al.* 1972) has shown that ALAD inhibition is similar in dogs and man, perhaps the rat observations are also similar to the effects in man. Lead and calcium have many similarities. It would be surprising if their absorption properties were vastly unalike. However, in order to show the effect of airborne lead in the presence of a low-calcium diet (as in children from poverty-level families), a controlled experiment should be set up using laboratory animals in which the effects of different levels of airborne lead with and without essential minerals is measured. As all animals would have the same dietary lead, the results should indicate the type of epidemiologic study to be followed for children.

The EPA position paper (Section VI, p. 1) admits that "it is unclear whether increased air lead exposures in urban areas are solely responsible for blood lead elevations." This point must be clarified by immediate experimentation and surveys as outlined in various places in this section. This should be done before the imposition of any national ambient air quality standard for lead.

The data used by EPA supporting the postulate (Section VI, pp. 2–4) that abnormal blood lead in adults is found predominately in urban areas is not totally convincing as the subject populations were not matched for all the other factors which might influence blood lead levels: food, water, prior exposure, existing total body lead burden, activity level, etc. No statistics were presented supporting the claimed percentage differences on these populations having blood lead levels above certain values. The results of the studies with children presented are confused by residence type, family income, etc. No way to separate the air lead effect from the total environmental effect on the children has been shown by EPA.

The EPA findings (No. 2, Section VII, p. 1) that blood lead (at concentrations below 40 μg/100 g of whole blood) inhibits ALAD function was not shown by the discussion in Section III. However, all available scientific evidence shows this to be true. The actual significance of this inhibition in affecting the mental health of children is unknown as yet, pending further epidemiologic studies as recommended above. EPA has not offered any demonstrable proof that cytochrome synthesis can directly and by itself be affected by airborne lead alone.

Finding no. 5 (Section VII, p. 3) is not supported by the evidence advanced by EPA. The effect that an air lead level of two $\mu g/m^3$ has on blood lead has not been shown due to the faulty interpretation of significant experiments and the use of faulty experimental results as discussed above.

EPA's conclusion regarding adverse response to questions in the *Federal Register* [37 (115), 11786-11787 (14 June 1972)] are presented in Appendix A of the EPA position paper and are generally sound and well supported by available data. In particular, EPA's decision (Appendix A, pp. 1–5) not to use the Goldsmith-Hexter regression (GOLDSMITH and HEXTER 1967) is proper from a statistical point of view despite other comments (KEHOE 1968, GOLDSMITH and HEXTER 1968). The review of the first EPA position paper (GOLDWATER 1972) is particularly convincing in this regard.

Blood lead is not a perfect index of recent lead exposure. In the absence of a more definitive measure, its use is reasonable as proposed by EPA (Appendix A, pp. 6–8).

The EPA estimates of the ranges in ventilation rate for adults and the general adult absorption factor for airborne lead are reasonable, the latter possibly being somewhat high. These estimates should not be used lightly as even moderate exercise can raise the ventilation rate by up to a factor of 12 and lead deposition and absorption in the lungs depend on particle size, shape, type, specific gravity, etc.

The supposition (Appendix A, pp. 18–26) that decreases in atmospheric lead emissions from mobile sources will cut down on lead in dust fall is reasonable. However, no immediate effect on the lead content of food and water is foreseen. As noted elsewhere in this analysis, the lead emissions from stationary sources must also be controlled. Very little of the nation's food supply is grown in fields immediately adjacent to high traffic density roads. Therefore, to decrease lead content in food requires that the total amount of lead in the environment must be decreased. This will be a very long process, and must be initiated now. If we wait for serious lead-related disorders to appear in the general population we may have passed the point of no return. The isotope ratio technique of CHOW (1970) should certainly be used to evaluate the sources of lead in all foods and public water supplies.

In this regard (Appendix A, pp. 29–30), a decrease in airborne lead should cause a corresponding decrease in lead in precipitation and slow down the gradual contamination of our soil and water. Again, all sources of lead emissions, not just automotive emissions, must be controlled for this effect to occur.

V. Conclusions II: Validity of scientific basis of proposed two $\mu g/m^3$ standard

From the foregoing analysis of EPA's position on the human health hazards of airborne lead and of the data used by EPA to support its decision, the following conclusions are evident: (1) factors other than vehicular emissions may be responsible for the high lead level in street dust; (2) control of airborne lead emissions as predicted by inhibition of δ-aminolevulinic acid dehydrase in man must depend on *in vivo* demonstration of this inhibition; (3) no valid epidemiologic data are available relating blood or urine lead levels to low (*ca.* two $\mu g/m^3$) lead levels in the ambient atmosphere; (4) a nationwide lead emissions inventory based on emission sampling and corresponding aerometric and epidemiologic study is required to determine the existance of an air lead vs. blood lead relationship; (5) any epidemiologic studies must consider all forms and sources of lead exposure and must not be based on physiological estimates and air lead and blood lead measures alone; (6) EPA has misused data from the "Seven City Study" and the "Roadway Study"; (7) the "Prisoner Study" did not provide any valid results to support the contention of human harm from low level (*ca.* 3 $\mu g/m^3$) lead exposures; and (8) there is no scientific basis for the proposed two $\mu g/m^3$ ambient air standard for lead.

Summary

A discussion of the current and potential human health effects of the steadily increasing emission of lead into the atmosphere is presented. The mechanism of the biochemical involvement of lead in the human body is through interference with the formation and degradation of hemoglobin and erythrocytes and by depression of the central nervous system. The best measure of these effects for short-term lead exposures is the lead content in blood. Unfortunately, no epidemiological data are available showing an unambiguous relation between elevated blood lead levels, ambient air lead concentrations, and frequency of lead-related disorders or diseases.

The scientific basis for the position taken by the U.S. Environmental Protection Agency in support of a national ambient air quality standard of two μg of lead/m^3 is also examined in detail. This position has been based on data that are incomplete, in some cases improperly obtained, in others incorrectly interpreted regarding the effects of low levels of atmospheric lead on human health. The EPA position is untenable because the available studies did not include all forms and sources of human lead exposure. Thus it is concluded that while further accumulation of lead in the ambient environment is unacceptable, there is no scientific basis for the proposed standard.

References

ALBAHARY, C.: Lead and hemopoiesis—The mechanism and consequences of the erythropathy of occupational lead poisoning. Amer. J. Med. **52**, 367 (1972).

ANONYMOUS: Lead in air (editorial). Brit. Med. J. **3**, 653 (1971).

AZAR, A., R. SNEE, and K. HABIBI: Relationship of community levels of air lead and indices of lead absorption. Presented Internat. Symp. on Environmental Health Aspects of Lead, Amsterdam, Oct. 2–6 (1972).

BAZELL, R. J.: Lead poisoning: Zoo animals may be the first victims. Science **173**, 130 (1971).

DE BRUIN, A.: Certain biological effects of lead upon the animal organism. Arch. Environ. Health **23**, 249 (1971).

CANNON, H. L., and J. M. BOWLES: Contamination of vegetation with tetraethyl lead. Science **137**, 765 (1962).

CHISHOLM, J. J.: Statement before the DPI *ad hoc* committee meeting, Cincinnati. Quoted from B. G. King in Amer. J. Dis. Child. **122**, 337 (1971).

CHOW, T. J.: Lead accumulation in roadside soil and grass. Nature **225**, 295 (1970).

——, and J. L. EARL: Lead and uranium in Pennsylvania anthracite. Chem. Geol. **6**, 43 (1970 a).

—— —— Lead aerosols in the atmosphere: Increasing concentrations. Science **169**, 577 (1970 b).

—— —— Lead isotopes in North American coals. Science **176**, 510 (1972).

——, and M. S. JOHNSTONE: Lead isotopes in gasoline and aerosols of Los Angeles basin, California. Science **147**, 502 (1965).

——, and C. C. PATTERSON: The occurrence and significance of lead isotopes in pelagic sediments. Geochim. Cosmochim. Acta **26**, 263 (1962).

DAINES, R. H., D. W. SMITH, A. F. FELICIANO, and J. R. TROUT: Air levels of lead inside and outside homes. J. Ind. Med. **41**, 26 (1972).

FELTON, J. S., E. KAHN, B. SALICK, F. C. VAN NATTA, and M. W. WHITEHOUSE: Heavy metal poisoning: Mercury and lead. Ann. Int. Med. **76**, 779 (1972).

GOLDSMITH, J. R.: Epidemiological bases for possible air quality criteria for lead. J. Air Pollut. Control Assoc. **19**, 714 (1969).

——, and A. C. HEXTER: Respiratory exposure to lead: Epidemiological and experimental dose-response relationships. Science **158**, 132 (1967).

—— —— Comment on lead intake from food and from the atmosphere. Science **159**, 1000 (1968).

GOLDWATER, L. J.: An assessment of the scientific justification for establishing 2 $\mu g/m^3$ as the maximum safe level for airborne lead. Ind. Med. **41**, 13 (1972).

——, and A. W. HOOVER: An international study of normal levels of lead in blood and urine. Arch. Environ. Health **15**, 60 (1967).

GRUNDY, R. D.: Session summary for carbon monoxide and lead. Part 3 of toxicologic and epidemiologic bases for air quality criteria. J. Air Pollut. Control Assoc. **19**, 729 (1969).

GUYTON, A. C.: Textbook of medical physiology, 3rd ed., p. 552. Philadelphia: Saunders (1966).

HAGGERTY, C. J., and R. A. NORLAND: An inquiry into certain aspects of lead poisoning in children as a community problem. Dept. Pediatrics Rept., Marquette Medical School (1969) as cited by B. G. King, Amer. J. Dis. Child. **122**, 337 (1971).

HALL, S. K.: Lead pollution and poisoning. Environ. Sci. Technol. **6**, 31 (1972).

HARDY, H. L.: Discussion of toxicological appraisal of lead in relation to the tolerable concentration in the ambient air. J. Air Pollut. Control Assoc. **19**, 701 (1969).

——, R. I. CHAMBERLIN, C. C. MALOOF, G. W. BOYLEN, JR., and M. C. HOWELL: Lead as an environmental poison. Clin. Pharm. Therap. **12**, 982 (1971).

HIRSCHLER, D. A., and L. F. GILBERT: Nature of lead in automobile exhaust gas. Arch. Environ. Health **8**, 109 (1964).

HUNT, W. R., C. PINKERTON, O. MCNULTY, and J. CREASON: A study in trace element pollution of air in 77 midwest cities. Presented 4th Ann. Conf. on Trace Substances in Environmental Health, Columbia, Mo. (1970).

KAHN, E.: Testimony presented at Los Angeles public hearing, p. 163 of hearing record, May 2–4 (1972).

KEHOE, R. A.: The Harben Lectures, 1960: The metabolism of lead in man in health and disease. J. Roy. Inst. Public Health **24**, 177 (1961).

—— Metabolism of lead under abnormal conditions. Arch. Environ. Health **8**, 47 (1964).

—— Lead intake from food and from the atmosphere. Science **159**, 1000 (1968).

—— Toxicological appraisal of lead in relation to the tolerable concentration in the ambient air. J. Air Pollut. Control Assoc. **19**, 690 (1969).

——, J. CHOLAK, and R. V. STORY: A spectrographical study of the normal concentrations of metals in biological materials. J. Nutr. **19**, 579 (1940).

KING, B. G.: Maximum daily intake of lead without excessive body lead-burden in children. Amer. J. Dis. Child. **122**, 337 (1971).

KOEPPE, D. E., and R. J. MILLER: Lead effects on corn mitochondrial respiration. Science **167**, 1376 (1970).

KNELSON, J. H., F. COULSTON, L. GOLDBERG, T. GRIFFIN, and R. J. JOHNSON: Kinetics of respiratory lead uptake in humans. Presented Internat. Symp. on Environmental Health Aspects of Lead, Amsterdam, Oct. 2–6 (1972).

LAGERWERFF, J. V., and A. W. SPECHT: Contamination of roadside soil and vegetation with cadmium, nickel, lead, and zinc. Environ. Sci. Technol. **4**, 583 (1970).

LANDAU, L., R. SMITH, and D. A. LYNN: Carbon monoxide and lead—an environmental appraisal. J. Air Pollut. Control Assoc. **19**, 684 (1969).

LAZRUS, A. L., E. LORANGE, and A. P. LODGE: Lead and other metal ions in United States precipitation. Environ. Sci. Technol. **4**, 55 (1970).

LEE, J. A.: Lead pollution from a factory manufacturing anti-knock compounds. Nature **238**, 165 (1972).

MCCAULDIN, R. O.: Estimation of sources of atmospheric lead and measured atmospheric lead levels. Symp. on Environmental Lead Contamination, *U.S. Public Health Service* Pub. No. 1440, pp. 7–16. (1966).

MCINTIRE, M., and C. R. ANGLE: Air lead: Relation to lead in blood of black school children deficient in glucose-6-phosphate dehydrogenase. Science **177**, 520 (1972).

MORSE, R. S., and G. A. WELFORD: Dietary intake of ^{210}Pb. Health Physics **21**, 53 (1971).

MUROZUMI, M., T. J. CHOW, and C. PATTERSON: Chemical concentrations of pollutant lead aerosols, terrestrial dusts, and sea salts in Greenland and Antarctic snow strata. Geochim. Cosmochim. Acta **33**, 1247 (1969).

OSTLE, B.: Statistics in research, 2nd ed., pp. 136, 201–205. Ames: Iowa State University Press (1963).

REDDICK, L. P.: Plumbism exists today. Southern Med. J. **64**, 446 (1971).

SCHMIDT-NIELSEN, K.: Animal physiology, 3rd ed., p. 22. Englewood Cliffs, N.J.: Prentice-Hall (1970).

SCHMITT, N., G. BROWN, E. L. DEVLIN, A. A. LARSEN, E. D. MCCAUSLAND, and J. M. SAVILLE: Lead poisoning in horses: An environmental health hazard. Arch. Environ. Health **23**, 185 (1971).

SCHROEDER, H. A., and I. H. TIPTON: The human body burden of lead. Arch. Environ. Health **17**, 965 (1968).

SINGER, M. J., and L. HANSON: Lead accumulation in soils near highways in the Twin Cities metropolitan area. Soil Sci. Soc. Amer. Proc. **33**, 152 (1969).

SMITH, W. H.: Lead and mercury burden of urban woody plants. Science **176**, 1237 (1972).

STANKOVIĆ, M. K.: Biochemical tests for the appraisal of exposure to lead. Arch. Environ. Health **23**, 265 (1971).

STEINFELD, J. L.: Surgeon General's policy statement on medical aspects of childhood lead poisoning (1970) as cited by B. G. King in Amer. J. Dis. Child. **122**, 337 (1971).

STOPPS, G. J.: Lead concentrations in blood and urine of "normal" populations. Presented Symp. on Environmental Lead Contamination, Washington, D.C. (1965).

—— Discussion of epidemiological bases for possible air quality criteria for lead. J. Air Pollut. Control Assoc. **19**, 719 (1969).

TATSUMOTO, M., and C. C. PATTERSON: The concentration of common lead in sea water. In J. Geiss and E. Goldberg (eds.): Earth science and meteoritics. Amsterdam: North Holland Publishing Co. (1963 a).

—— —— Concentrations of common lead in some Atlantic and Mediterranean waters and in snow. Nature **199**, 350 (1963 b).

TEPPER, L., and L. LEVINE: A survey of air and population lead levels in selected American communities. Final report, Contract PH 22-68-28, *U.S. Environmental Protection Agency* (1972).

TER HAAR, G. L., and M. A. BAYARD: Composition of airborne lead particles. Nature **232**, 553 (1971).

——, D. L. LENANE, J. N. HU, and M. BRANDT: Composition, size and control of automotive exhaust particulates. J. Air Pollut. Control Assoc. **22**, 39 (1972).

DE TREVILLE, R. T. P.: Natural occurrence of lead. Arch Environ. Health **8**, 212 (1964).

U.S. Department of Health, Education, and Welfare: Air quality data from the National Air Surveillance Networks and contributing state and local networks for 1964–1965, pp. 82–84. Cincinnati, Ohio (1966).

—— Air quality data from the National Air Surveillance Networks and contributing state and local networks, 1966 ed. Rept. No. APTD 68–9, pp. 100–102. Durham, N.C. (1968).

U.S. Environmental Protection Agency: Air quality data for 1967 from the National Air Surveillance Networks and contributing state and local networks, rev. 1971. Rept. No. APTD-0741, pp. 109–110. Research Triangle Park, N.C. (1971).

—— Air quality data for 1968 from the National Air Surveillance Networks and contributing state and local networks. Rept. No. APTD-0978, pp. 124–127. Research Triangle Park, N.C. (1972 a).

—— EPA's position on the health effects of airborne lead. Washington, D.C. (1972 b).

—— OAP data file of nationwide emissions 1970, table G.2: Lead emissions in the United States, 1970. Durham, N.C. (1972 c).

ZIELHUIS, R. L.: Interrelationship of biochemical responses to the absorption of inorganic lead. Arch. Environ. Health 23, 299 (1971).

ZOOK, B. C., J. L. CARPENTER, and R. M. ROBERTS: Lead poisoning in dogs: Occurrence, source, clinical pathology, and electroencephalography. Amer. J. Vet. Res. 33, 899 (1972).

Manuscript received December 10, 1973; accepted January 22, 1974.

Lead and plants

By

Wolfgang Höll and Rüdiger Hampp*

Contents

I. Introduction

Because of its low melting point, its malleability, and its inertness, lead has been a favored metal of man since pre-Christian times. With progressing civilization lead has been mined in increasing amounts from natural sources and has been partly returned into the environment after its technological use.

Besides the contamination of the ecosystem by man, there are natural lead sources. Natural concentrations have been estimated to be about 0.0005 μg/cu m of air (Patterson 1965). The natural lead concentration of air is caused by erosion of lead-containing soils (Chow and Patterson 1962). A second natural source of lead is gases which are exhausted from the earth's crust (Blanchard 1966). Natural lead contents of the atmosphere apparently contribute only insignificantly to present-day concentrations in the air. Investigations of chronological layers of snow strata of the polar region (Murozumi *et al.* 1969) demonstrated very impressively that with the beginning

* Department of Botany, Technical University, 8000 München 2, Arcisstr. 21, Germany.

of the Industrial Revolution (about 1750) there has been an increase of emitted lead. This increase of lead pollution in the 18th and 19th centuries is mainly due to smelting processes and the combustion of coal. With the improvement of smelting procedures and the modernization of coal-furnaces, the amount of lead pollution from these industries could be lowered in our century.

The introduction of lead compounds as anti-knock substances in gasolines in the twenties of this century established a new and potent lead source, dangerous for the environment. The distinct rise of lead concentration in air, water, and soil since about 1950 has been caused by the tremendous increase in use of these leaded gasolines (CHOW and EARL 1970). This problem of lead exhaust from cars has received great consideration in recent years (see literature cited by DANIELSON 1970; Report of the *Eidgenössiche Kommission für Lufthygiene* 1971; Report of the *Committee on Biological Effects of Atmospheric Pollutants* 1972). According to ATKINS (1969), a car driving 90 km/h exhausted about two mg of lead/sec. Direct evidence that the overwhelming part of the lead in the environment derives from gasoline combustion was obtained by investigations on isotopes (CHOW 1970). Lead deposited on plant surfaces exhibited the isotope combination $Pb^{206}/^{203} = 18.41$, $Pb^{206}/^{207} = 1.197$, $Pb^{206}/^{208} = 0.4827$, and is thus identical with the isotope combination of lead in gasoline.

The Air Pollution Control Office in the United States has provided for 1968 a data bank that shows that approximately 98% of the lead in the environment derives from the combustion of leaded gasoline. During the process of combustion, lead is released into the atmosphere in the form of aerosols. Particle size of the lead aerosols is an important parameter in the discussion of the distribution of lead from its emission source. A number of investigations (MUELLER 1970, ROBINSON *et al.* 1963, BÜTTNER 1972, LAVESKOG 1970) demonstrated that 50 to 90% of the lead particles are less than one μm in diameter. Studies with ^{210}Pb showed that lead of this particle size remained suspended in the air up to 30 days (FRANCIS *et al.* 1970). This time is sufficient to distribute lead worldwide. There is a difference in the nature of lead compounds between bigger (diameter between two and ten μm) and smaller lead particles. The bigger lead particles present the metal mainly as bromochloride (PbCl·Br), whereas the smaller particles (less than one μm) consist of *alpha*- and *beta*-forms of ammonium chloride and lead bromochloride ($NH_4Cl \cdot 2PbCl \cdot Br$, 2 $NH_4Cl \cdot PbCl \cdot Br$), small amounts of lead sulfate ($PbSO_4$), and the mixed halogen oxide ($PbO \cdot PbCl \cdot Br \cdot H_2O$) (see HIRSCHLER and GILBERT 1964).

The lead compounds which are exhausted by combustion engines are concerted in the atmosphere, which is not a natural photochemical process: about 18 h after emission 30 to 40% of the chlorides and 75% of the bromides were transformed into carbonates and oxides (TER HAAR and BAYARD 1971; see also DALE and WINCHESTER 1972, DALE

1972). The aerosol-bound lead in the atmosphere is precipitated by rainfall. TER HAAR *et al.* (1967) showed that the intensity of rainfall was of importance for the lead sedimentation: a gush of rain was less effective in lead precipitation than a constant slight rain.

In addition to the lead in the atmosphere, the concentration of this heavy metal in the soil is of importance to plant life (TSO *et al.* 1966). It has been known for a long time that lead influences the soil environment (LIPMAN and BURGESS 1914). Interestingly, the top soil particles accumulated most of the lead (SWAINE and MITCHELL 1960, RÜHLING and TYLOR 1968, LAGERWERFF and SPECHT 1970, PAGE and GANJE 1970). With increasing soil depth, the lead content is lowered significantly. Humus soils are able to bind lead more strongly than other soils. It is possible that lead and humus compounds form organometallic complexes (HASSLER 1943) which are much more stable than only absorption of lead onto minerals of the soil.

II. Lead in plants

Publications concerning lead concentrations in plant tissues can be summarized under the following aspects.

a) Correlation between lead content and intensity of lead pollution

In nearly all cases there existed a positive correlation between lead content and the intensity of lead pollution (see literature cited by GARBER 1973).

CANNON and BOWLES (1962) analyzed specimens of grasses grown at different distances (one to 1,000 ft) from a street and found up to 3,000 ppm of lead (calculated on ash weight basis) in the direct vicinity of the roads. Vegetables grown in gardens near streets (<50 ft) contained an average of 115 ppm (values ranged from ten to 700 ppm) of lead. With increasing distance from the pollution source the lead contents of the specimens investigated decreased. Similar results were obtained by KLOKE and RIEBARTSCH (1964). Plants grown in areas with low traffic intensity (11,000 cars/12h) contained an average of 16.2 ppm of lead/g dry weight. The lead contents increased to 57 ppm at areas with higher traffic load (32,000 cars/12h). WARREN and DELAVAULT (1962) reported that one-year-old trunks of trees near high traffic areas contained much more lead than trees of the same age but far away from streets. Thus, the wood of ash, oak, and hazel in the City of London contained 14, 20, and 52 ppm of lead, while trees grown in Sussex, a rural area, showed only 2.0, 0.8, and 2.0 ppm of lead, respectively. Investigations performed by the same authors on field crops showed that at a distance of 275 m from roads with high traffic intensity, lead concentrations were below ten ppm.

LEH (1966) analyzed rice plants, grown at a distance of five m from a highway. The grains contained 5.2 ppm whereas the trash contained 205 ppm of lead. At a distance of 92 m from the highway the values decreased to 4.8 and 11.8 ppm, respectively. At the same locations the lead content of the shoots of potato plants decreased from 40 to 21 ppm, while that of the tubers exhibited only insignificant changes.

Analyzing wheat and barley, SUCHODOLLER (1967) found high lead concentrations only in the direct vicinity of a street, with up to a maximum of 147 ppm of lead/g dry weight. With increasing distance, a pronounced decrease of lead content was established: after 30 to 40 m no influence from the street was detectable. Near roads with high traffic intensity, EVERETT *et al.* (1967) found in the leaves of privet plants 86 ppm of lead, while those in the vicinity of streets with only little traffic load contained 45 ppm (see also SINGER and HANSON 1969).

From the toxicological standpoint, investigations performed by TER HAAR (1970) are of interest. He investigated the edible parts of tomatoes, potatoes, wheat, carrots, lettuce, and oats grown near US Highway No. 24 (29,000 cars/day) in the neighbourhood of Detroit. There was no relation between lead contents in the edible parts of the plants and distance (nine to 159 m) from the highway; the vegetative parts, however, showed a positive correlation. Similar results were obtained by BAUMHARDT (1970) with maize plants; even at very high lead concentrations of the soil, no lead could be detected within the grains. On the other hand, MOTTO *et al.* (1970) reported an increase of lead contents in turnips, wheat, lettuce, potatoes, and tomatoes with increasing traffic intensity. These investigators demonstrated that the maximum lead concentration was limited within a zone of 31 m from the highway. In the same way, PAGE *et al.* (1971) analyzed the lead contents of 27 crop varieties in different distances to highways in southern California. They also were able to demonstrate a positive correlation between the distance from the street and the amounts of lead in the crops. Exceeding the distance of 150 m no direct relation was measurable. The lead content in this distance seemed to be controlled by regional and not by local parameters. Within a distance of 50 m to the border of the streets a maximum lead content in plant tissues was measured.

In honey collected in different localities, KERIN and KERIN (1971) found the following lead concentrations: influenced by lead industry = 7.80 ppm, uninfluenced towns = 0.46 ppm, and uninfluenced villages = 0.18 ppm.

ROHMEDER and WEBER (1970) investigated the influence of flight traffic on the lead accumulation in spruce tree needles, in a wood near the Munich-Riem airport. The wood was crossed 240 times/day by planes during the years 1968 to 1970. One-year-old blighted needles contained 134.7 ppm of lead/g dry weight, while needles two years of age exhibited a lead content of 128.5 ppm. According to these authors, the lower lead content in the two-year-old needles was due to the adoption of lead-free jet fuels. Normal needles exhibited average lead contents of 14.7 (one-year old) and 30.8 ppm (two-years old). In the vicinity of a lead smelting furnace, D. KERIN (1971) found 30 to 500 ppm of lead in the needles of pines and spruces.

The contamination of vegetables with lead has been analyzed by several investigators (see Z. KERIN 1968, DJURIC *et al.* 1972, ZUBER 1972). FLETCHER (1971) reported lead contents as ppm of lead (dry-weight basis) of the following crops, washed before analysis: corn 2.7, bean 3.4, cabbages, 1.9, tomatoes 2.7, turnips 3.5, cucumbers 5.6, lettuce 17.4, and potatoes 6.6 and 7.5.

GARBER (1970) investigated the influence on plants in the vicinity of industries utilizing heavy metals. He analyzed the concentrations of some heavy metals in rainwater collected at different distances from the factories. Concerning lead, the following data were obtained: "near the factory" 66.1 mg/l, at a distance of 1,100 m 33 mg/l, and "far away" 0.04 mg/1. In the direct neighborhood of the factory pine needles contained 1,826 ppm, grass leaves 1094 ppm, and poplar leaves 398 ppm of lead.

HEILENZ (1970) analyzed fodder and food-stuff plants and found in districts far away from the influence of both traffic and industries one to six ppm of lead (dry-weight basis). Specimens of grasses from areas influenced by highways exhibited the following lead concentrations: dividing lane 64.6 ppm, border of the highway 49.6 ppm, at five m 34.5 ppm, at ten m 16.4 ppm at 20 m 16.1 ppm, at 50 m 6.3 ppm, and at 100 m 3.7 ppm. The lead contents of potato tubers and grain kernels, however, exhibited no correlation to distance from the highway.

NILSSON (1972) investigated spruce needles and trash in three different areas of southern Sweden. Although the distances from roads exceeded 300 m, a distinct lead accumulation was established in all areas. Depending on age and environmental conditions, the lead content of the needles was between 0.8 and 7.1 ppm (dry-weight basis). In common, the older needles exhibited a higher lead content. For the trash, values from 26.8 to 105.0 ppm were obtained. Extremely high lead contents were found by KELLER and PREIS (1967) in spruce needles grown in areas in Switzerland with high traffic load, with more than 600 ppm (dry-weight basis).

Individual plant genera often exhibit pronounced differences in their ability to accumulate lead. According to PALUCH and KARWETA (1968 and 1970), white clover as compared with other plants in the vicinity of a metallurgical industry showed the highest ability to accumulate lead. HOPKINSON *et al.* (1972) demonstrated that plants grown at different distances from a highway two miles south of Austin, Texas contained varying amounts of lead. The highest lead concentrations (dry-weight basis) were found in *Phyllantus polygonoides* (9,100 ppm), *Sporobolus spec.* (1,700 ppm), and *Sida filicaulis* (1,100 ppm).

In comparison to terrestrial plants, epiphytic forms accumulate to a higher extent substances taken up via the surface. This behavior is reflected also by concentrations of heavy metals in the tissue. GRUPE (see GARBER 1973) introduced the lichen *Parmelia physodes* as a bioindicator for the intensity of lead emission. At a distance of 1,000 m from a highway the thalli contained 35 ppm (unwashed) and 19 ppm (washed), respectively, of lead.

MARTINEZ *et al.* (1971) demonstrated that the lead content of *Tillandsia recurvata* L. varied with distance from the pollution source and traffic intensity. In an extreme case, lead represented 0.085% of the dry weight.

Investigations on lower plants, mainly mosses, confirm the results obtained on higher plants. LEE (1972) demonstrated that tissues of mosses exhibit lower lead contents with increasing distances from lead-using industries. At 800 m from the contamination source, the lead content was two to three times higher than in those mosses grown in uncontaminated areas. In specimens of *Marchantia* obtained from different areas (Edinburgh and Glasgow), BRIGGS (1972) found lead values, even at great distances from roads, which were due to contamination. The lead contents corresponded to the respective traffic intensity and were without significant correlation to the lead content of the soil.

b) Influence of wind direction

DUNN and BLOXAM (1933) reported the influence of horizontal wind-movement on the distribution of atmospheric dusts. The lead content of grasses grown leeward of a highway was significantly higher than that in plants on the windward side. CANNON and BOWLES (1962) demonstrated that grasses on the eastern side of U.S. Highway No. 287, running from north to south, contained 50 ppm of lead up to a distance of 100 ft (dry-weight basis). This value was much less in plants on the western or windward side of the highway even after ten feet. SUCHODOLLER (1967) found in plants grown leeward to a street decreasing lead gradients with increasing distance to the street, whereas on the windward side a discontinuous distribution without any relation to the distance from the street was measurable. Lead values on the leeward side were pronouncedly higher. Similar results have been reported by other authors (HAMMOND and ARONSON 1964, PAGE *et al.* 1971, CHOW 1970, FIDORA 1972, HAMPP 1973 a) (see Fig. 1).

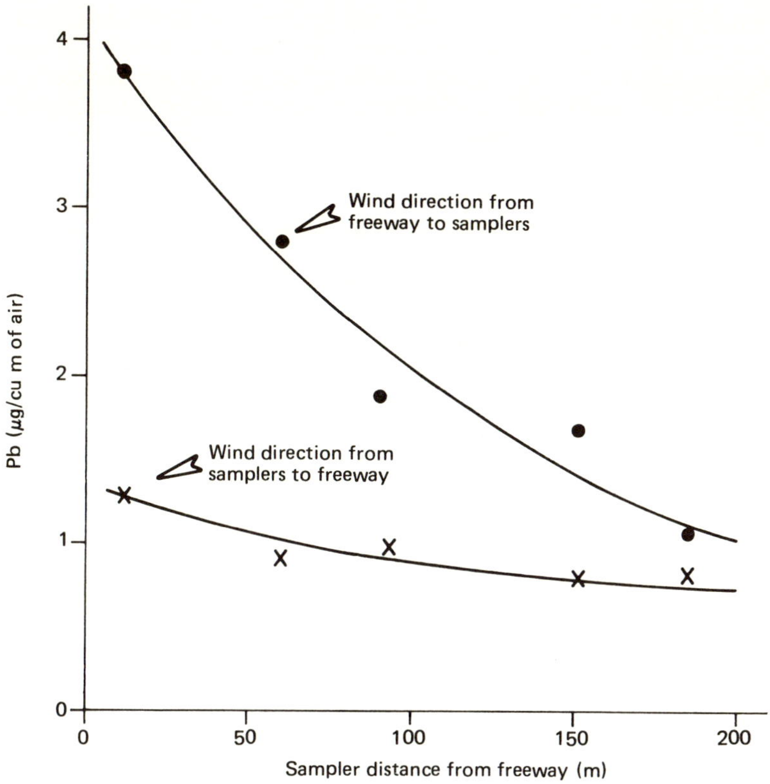

Fig. 1. Average ambient lead concentrations in a field plot on the Irvine Ranch as influenced by sampler distance and wind direction from Santa Ana freeway (from PAGE *et al.* 1971).

c) *Removable part by water wash*

Concerning the percentage of lead deposited at/in plants which can be removed by water wash, the reports are very much at variance. Thus, KLOKE and RIEBARTSCH (1964), investigating grasses, reported that a very large degree (80% and more) of the lead was washed off by water. In contrast to this statement, GARBER (1970) found only between 30 and 50% of the total lead was removable by water washes. SUCHODOLLER (1967) demonstrated a correlation between the removable part and the total lead content (see Fig. 2). If there was a high lead content, a relatively large part was removed by a water wash (up to 50% in the first three hours). In the presence of low lead contamination, proportionally minor quantities were removable. This result points towards an absorption, which is saturated at a definite concentration.

One can state that lead is partially deposited on the surface and partially incorporated into the tissue because the total lead content showed a correlation to the leaf area, whereas after a washing period of 24 h the lead content still was a function of leaf weight. PAGE *et al.* (1971) and FIDORA (1972) demonstrated that the surface structure of a plant is of importance for ability to accumulate lead. Plant tissue with a rough or hairy surface structure accumulated significantly more lead within a given time at the same location than those plants with a smooth surface. In general, lead retention ability is higher in plants with a rough or hairy surface (see also QUINCHE *et al.* 1969, KLOKE and LEH 1969).

Mild washing procedures suffice to remove a high percentage of lead from plant parts with a smooth surface, whereas in the case of rough surfaces intensive treatment is necessary to remove the same amount. The results of RAINS (1971) point toward a relatively strong binding of lead onto plant tissue surfaces. Using water which contained EDTA, significantly more lead was removable than when using 0.01N HCl or water. Even with ammonium acetate solution, only insignificant amounts of lead were removed from *Tillandsia recurvata* L. (MARTINEZ *et al.* 1971).

The data available till now indicate that the accumulation of lead

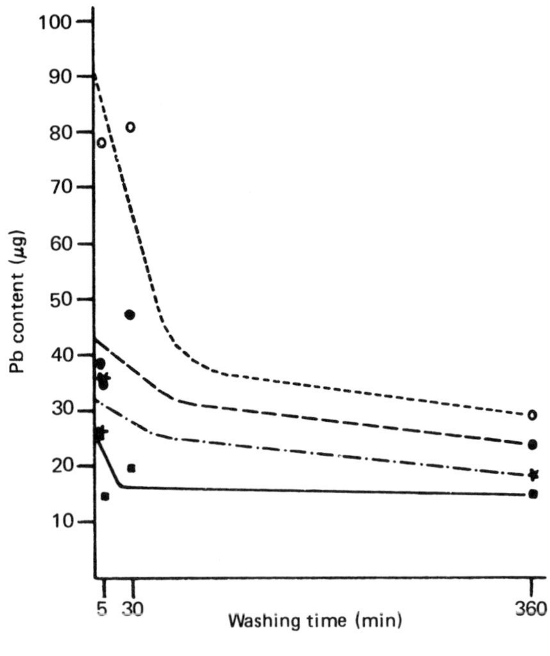

Fig. 2. Removable part of lead by water wash in grasses grown at different distances from a street (from SUCHODOLLER 1967); distance (m): 0-----, 10·-·-·, 5---, and 30——.

by plants depends on the surface structure. That part of the lead which can be removed by washing with water, therefore, is no indication of the true contamination.

d) Distribution in different plant organs

One can expect that lead which is supended in the air is accumulated in those plant organs which exhibit large surface area and intensive gas exchange with the surrounding atmosphere.

SUCHODOLLER (1967) demonstrated that the leaves of cereals grown near highways exhibited a maximum of 147 ppm (dry-weight basis), whereas the spikes contained 48 ppm and the stems 20 ppm of lead, respectively (see also LEH 1966, TER HAAR 1970, MOTTO *et al.* 1970). According to PAGE *et al.* (1971), the lead contents of plant tissues which are not in direct contact with the air, such as roots of legumes, are only insignificantly influenced by pollution of highways in the vicinity.

Analyzing *Acer platanoides* L. grown in the City of Munich, lamina of the leaves exhibited a higher lead content than fruits and leaf stalks (HAMPP 1973 a).

For the nutrition of man it is important that all protected parts of the plants, which in almost all instances are the edible portions, contain only very low lead levels which show no correlation to the lead load of the location.

e) Seasonal variations

Plants with dormant periods in their life cycles exhibit a fluctuation in lead content. One can expect that after spring butting, in the course of the vegetation period a constant rise of lead content in the overground parts occurs. PAGE *et al.* (1971) found that in the fruits of tomato plants a constant increase of lead content runs parallel to the ripening process.

In *Avena fatua,* RAINS (1971) discovered an annual fluctuation in lead content which exhibits dependencies on a constant supply from the atmosphere and a change of the relation lead quantity/dry weight. During the intensive growth period from April to June a relative decrease of this relation was measured, and it in turn increased with progressive ripening. The maximum content was measured in winter. According to this author, this phenomenon could be explained by different growth activities. Thus, during the enhanced growth period in spring the lead is diluted, for with progressive ripening a loss of matter (loss of dry weight) in the magnitude of 10 to 20% occurs. This relation between the relative lead content and the growth intensity was confirmed by FIDORA (1971). Plant material near highways, analyzed at different times of the vegetation period during springtime rapid growth, exhibited a more or less constant lead con-

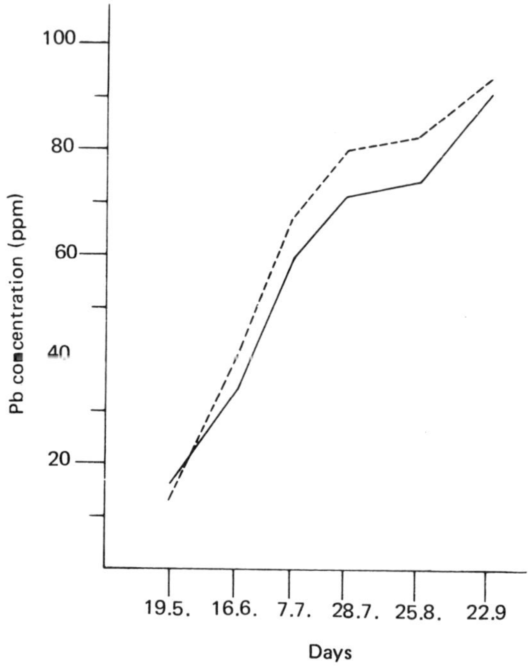

Fig. 3. Lead concentrations (dry weight) in leaves of *Betula verrucosa* Ehrh. grown at a distance of three m on the west (solid line) and east (dashed line) sides of a highway, varying during the vegetative period (from FIDORA 1972).

tent. During summertime however, a continuous accumulation was measured (see Fig. 3). At locations with very high traffic loads, saturation of the leaves with lead was established at the end of the vegetative period.

f) Cellular localization

HAMMETT (1928 a), treating thin sections of *Allium cepa, Phaseolus vulgaris,* and *Zea mays* with lead nitrate, demonstrated that maximum lead accumulation occurred in tissues with high mitotic activity. Especially along cell wall, crowded rows of crystals which seemed to be localized partly within the cell walls were observed (HAMMETT 1928 c). The supposition of HAMMETT, that the cell wall possesses a specific affinity towards lead, was confirmed by investigations on lower plants. On lichens (*Cladonia rangiformis* Hoffm.) of a lead mine population, washing experiments with various concentrations (10^{-6} to $1N$) of HCl resulted in a pronounced lead elution starting at a concentration of $10^{-3}N$. At the highest HCl concentration used, lead was removed quantitatively. With water, however, only insignificant amounts of lead were washed off. The results indicated that lead is obviously bound to insoluble anionic binding sides in an exchangeable form (see CLYMO 1963, TUOMINEN 1967, RÜHLING and TYLER 1970).

Since it is known that the metal tolerance of angiosperms shows a relation to the ionic binding of these metals to polyuronic acids of the cell wall (PETERSON 1969, TURNER and MARSHALL 1971 and 1972), it seems possible that lead also is bound in this form. HAMMETT (1928 c) demonstrated a significant lead accumulation in the nucleus of the cell. Treatment with sodium sulfide precipitated the lead present in the nucleus in the form of PbS and was thus identified. This result of lead accumulation in the nucleus was first confirmed by DALLENBACH (1964) and more recently by SKAAR *et al.* (1973). The mosses investigated by SKAAR *et al.* (grown at a distance of 250 m from a street with high traffic intensity) showed, in contrast to uncontaminated plants, the accumulation of electron-dense particles in their nuclei. Using the electron microscope microanalyzer, it was demonstrated that these particles consisted of lead compounds. The authors presumed that lead was bound within the nuclear membrane in the form of a nondiffusable complex. The enrichment within the nucleus should prevent a concentration of this metal toxic to enzyme systems located in the cytoplasmatic fraction.

Within isolated cell organelles such as mitochondria (YODAIKEN 1966) and chloroplasts (HAMPP 1973 b), lead often reaches extremely high concentrations. Chloroplasts isolated from contaminated spinach leaves contained up to 44% of the overall amount of lead (HAMPP 1973 b). The statements of a lead accumulation within the cell organelles point towards a correlation with the intensity of metabolic processes. In all cases a binding of lead to phosphate groups seem to predominate.

III. Lead tolerance

Within a plant species there are lead-tolerant and lead-sensitive breeds. BRADSHAW (1952) investigated the sensitivity of various populations of *Agrostis tenuis* Sibth. towards lead. Cultivating these lead-adapted plants, which grow in soils containing up to one % lead, in lead-free soils, a retardation of growth was observed. On the other hand, a high proportion of nonadapted seedlings died when cultivated in lead-containing soils. It is of interest that the lead-adapted *Agrostis tenuis* plants do not lose resistance towards high lead concentrations, even when they are cultivated for more than two years in lead-free soils. These findings point toward a lead resistance which seems to be genetically fixed.

WILKINS (1957) found that the lead resistance of plants which were cultivated in nutrient broth is extraordinarily low. Even at a lead nitrate concentration of three ppm an inhibition of root growth was observed. In the presence of Ca^{2+} the lead resistance was drastically raised. Test series performed by WILKINS (1957) showed that plants grown near a lead mine (*Agrostis tenuis, Deschampsia flexuosa, Festuca ovina*) exhibited a low sensitivity towards this heavy metal. JOWETT (1958) supposed that the pronounced insensitivity of lead-adapted plants is due to the presence of a tolerance mechanism. According to this author, lead-resistant plants are able to bind metal ions in the form of chelates. This binding ability seems to be specific for certain metals. If there exists a pronounced ability for complexing lead ions, other metals, for instance copper, are much less bound by the plants. The advantage for the plants may be that, even in the presence of high lead concentrations, micronutrients essential for growth remain soluble.

Later, JOWETT (1964) showed that lead resistant *Agrostis tenuis* populations exhibit an adaption on very low Ca^{2+} and PO_4^{3-} concentrations. This is necessary, because high lead contents of the soil lower phosphate concentrations essential for plants. Populations of a lead mine showed, in contrast to noncontaminated plants, no reaction towards different concentrations of Ca^{2+}. Mosses also reveal a pronounced lead tolerance. Investigations on clones of *Marchantia* (BRIGGS 1972) pointed towards a genetic fixation of the lead tolerance.

IV. Uptake and translocation

Field experiments by BERRY (1924) demonstrated that application of lead nitrate improved the fertility of soil. Even when relatively high lead nitrate concentrations were used, no significant accumulation of the heavy metal was measured within the plants. The melioration of the soil could be deduced without doubt to the nitrate ion (see also EHRENBERG and NOLTE 1916, SCHARRER and SCHROPP 1936).

In contrast to BERRY (1924), PRÁT (1927) demonstrated that the lead from a $10^{-4}M$ $PbCl_2$ solution was taken up by *Vicia faba* plants in less than three days. After removal of the cotyledons, a slowdown of lead uptake was measured. This effect is possibly due to the reduced transpiration stream. KEATON (1937) demonstrated that the upper parts of plants exhibited only low lead concentrations of from traces up to three ppm when the heavy metal was applied to the soil. The lead content in these plant parts showed no correlation with the applied metal concentrations. The roots, however, exhibited a positive correlation. Lead in the form of nitrate was taken up faster than lead in the form of carbonate. These findings were confirmed by JONES *et al.* (1973). They demonstrated in experiments with soil-grown rye grass that roots restricted the movement of lead into the tops of high-yielding plants. When growth, however, was limited by sulfur deficiency the concentration of lead in the tops increased markedly.

One of the first investigators to introduce a radioactive lead isotope in studying uptake of lead by plants was HEVESY (1923). According to him, 0.3% of the lead present in a 0.1*N* lead nitrate solution was taken up by *Vicia faba* plants within 24 h, whereas 60% of the lead was taken up when a $10^{-6}N$ solution was used. The lead concentration in the leaves increased only insignificantly during the test-period. Upon transferring the contaminated plants into an isotope-free lead salt solution, the label was exchanged with nonlabeled lead. This result indicates that there may be a reversible binding of lead in the form of a dissociable salt. Using autoradiographic methods, SUCHODOLLER (1967) also demonstrated a very less pronounced translocation into the stem. In bean plants, however, a much more rapid lead transport into organs far away from the roots was measured.

According to BAUMHARDT (1970), lead is taken up by maize plants in detectable quantities when the lead concentration in the soil is 250 μg/g. Even at very high lead concentrations in the soil, no lead was detected in the seeds (see also MOTTO *et al.* 1970). The different velocities in lead transport within plants seem to depend on the amount of lead which reaches the xylem strands. Microautoradiographic studies on barley showed a pronounced lead enrichment in the root cortex, whereas only traces of lead were found in the protoxylem region.

In common there is no linear correlation between lead translocation within plants and the increasing concentration in the nutrient solution. KELLER and ZUBER (1970) demonstrated that in needles of spruce trees (five years old) there was an increase in lead content of only 2.5 ppm even when the concentration of lead in the nutrient solution was increased from zero to 1,000 ppm. This very low lead accumulation in the needles is due possibly to the low transpiration velocity of conifers.

Lead uptake experiments, performed with different moisture pressure deficits, showed that there are only insignificant differences in the needle contamination of the spruce trees investigated. Feeding experiments performed with tomato, barley, and bean plants (BROYER *et al.* 1972) showed that roots of the respective plants contained one to two magnitudes of order more lead than the shoots. Z. KERIN (1970) found a positive correlation between the lead content of the soil and the lead content of carrots grown in it. SOMMER *et al.* (1971) obtained similar results on other plants (see also BOVAY 1970). LAGERWERFF (1971) demonstrated the influence of soil pH on lead uptake. At pH 5.9 the water-soluble part of the lead present in the soil was increased. At the same time a significant increase of lead in the shoot was measured. Increasing the soil pH from 5.9 to 7.2 the solubility and, accordingly, the motility of lead decreased (see also MACLEAN *et al.* 1969, ZUBER and BOVAY 1972).

A possible pathway of lead uptake is discussed by GOREN-

SUCHODOLLER and WANNER (1969). If one applies lead and manganese to intact roots of barley, both metals showed a two-phase uptake into the tissue. Most of the lead was taken up at very low concentrations ($10^{-4}M$). Raising the concentration in the medium, only little additional lead was accumulated (see Fig. 4). These phases of metal uptake seem to be physical in nature, because the temperature quotient is almost one. The authors cited showed that the lead uptake was stopped after one to two h. The lead which was taken up during this period was not removable by washing the root tissue with a solution of calcium nitrate. In contrast, manganese was washed off almost entirely. From the results that Mn^{2+} uptake was inhibited in the presence of Pb^{2+} and, on the other hand, the lead uptake was influenced by Mn^{2+}, GOREN-SUCHODOLLER and WANNER (1969) concluded that the absorption sites of manganese and lead ions were possibly identical. The affinity of lead ions to these sites was higher than that of manganese ions. The lead content in above-ground organs of plants grown in localities with high traffic load is due to the uptake

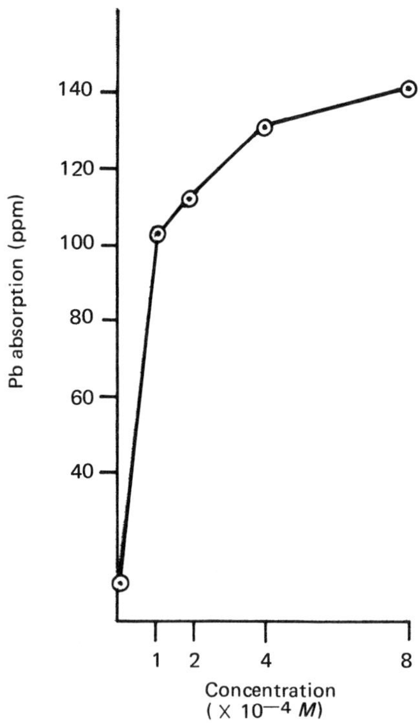

Fig. 4. Lead absorption by barley roots (intact plants, eight days old) in the presence of different concentrations of lead nitrate. Values for absorption are on a dry-weight basis (from GOREN-SUCHODOLLER and WANNER 1969).

by the leaves. This statement has been substantiated by numerous investigations (see TER HAAR 1970, GARBER 1970, TUNNEY *et al.* 1972).

The amount of lead translocated within the plant body can only be a subject of speculation. Some indications point towards the possibility that besides a translocation in apical direction there exists one in basal direction.

A probable lead transport in the xylem was shown by HEVESY (1923), SUCHODOLLER (1967), and Tso and FISENNE (1968). The translocation of lead occurred predominantly in the apoplast. In trunks of hardwoods a gradient in longitudinal direction was detected (HAMPP and HÖLL 1974). In the wood tissue (sap- and heartwood values combined), a decrease in lead content with increasing trunk height was detected. This finding points towards an apoplastic lead transport from the roots with the transpiration stream.

Investigations on young twigs of *Acer pseudoplatanoides* L. trees grown in areas with high traffic intensity indicated that there is also a phloem transport (HAMPP and HÖLL 1974) (see Fig. 5). With increasing distance from the leaf insertion points there was a decrease in lead content in the phloem tissue. The lead measured was probably taken up by the leaves, because the lead concentration of the tissue near the leaf insertion point exhibited a positive correlation with the leaf area above the point of measurement. The lead concentration below the insertion of two twigs was generally higher than that of

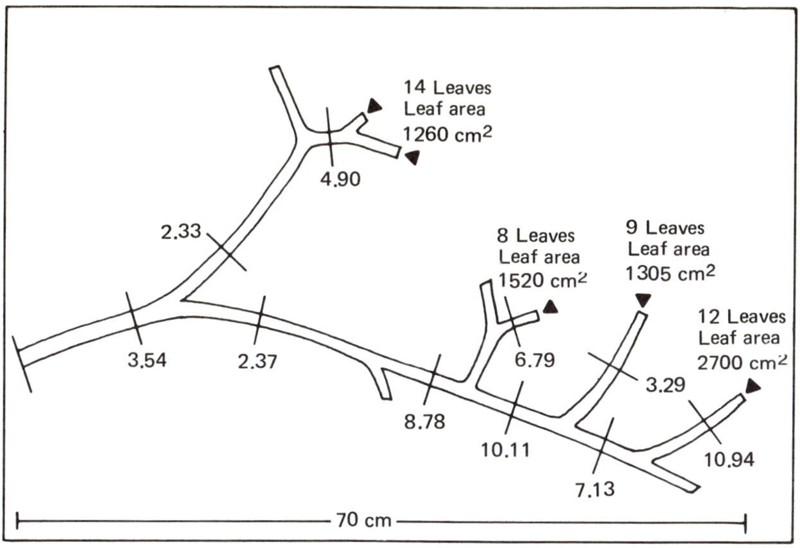

Fig. 5. Lead contents in different parts of a terminal twig of *Acer platanoides* L.; values represent µg of lead/g dry weight (from HAMPP and HÖLL 1974).

either twig directly above the intersection. These findings of a basal translocation of lead are in accordance with results of other investigators. MITCHELL and REITH (1966) found that the lead in leaves was translocated into the stem prior to winter dormancy. These results make a phloem transport probable. Because of the high concentration of free phosphate in the sieve tubes (ZIEGLER 1956, TAMMES and VAN DIE 1966), one would have to assume that the lead is present as a complex.

Chelates play a major role in uptake processes in plants. A number of elements which, due to the low solubilities of their salts, are taken up by the plants only in small amounts, are made available for the plants by forming ligand binding (see STEWART 1963). All the amino acids, which represent the main component transported in the phloem, are to a certain extent able to form chelates with metal ions. Histidine and cysteine for instance are powerful chelate formers (see CRAFTS and CRISP 1971, VALLEE and ULMER 1972). The high stability constants for 1:1 complexes between lead and histidine (6.84) and cysteine (12.2), respectively, make it highly possible that a lead transport in the phloem takes place even in the presence of high phosphate concentrations.

HAMPP and HÖLL (1974) indicated that a parenchymous translocation can also occur. In comparison to the heartwood, the sapwood of black locust trees exposed to high lead pollution exhibited higher lead concentrations. It is evident that the high lead content is associated with the living tissue of the trunk. From the findings of these authors it is clear that perhaps 70% of the lead which is present in the sapwood is fixed in the form of immobile compounds during the process of heartwood formation. The remaining 30%, however, seems to be mobile and is translocated in a centrifugal direction with the sap-heartwood boundary in each vegetative period. In trees which exhibited no pronounced heartwood formation and which possessed living parenchymous cells in each part of the trunk, a more-or-less statistical distribution of lead was demonstrated in cross sections. This possible translocation across the trunks is probably the reason why there is no correlation between lead contents of wood tissues differing in age and the increase of lead pollution within the last few years.

V. Physiology

At the beginning of this century attempts were made to increase crop yields by applications of heavy metals such as iron, manganese, and lead. STOKLASA (1914) increased the yields of barley, corn, and buckwheat to 90% by applying lead nitrate. The optimum increase was obtained when 2.7 to 4.3 kg of lead nitrate were applied/A.

Similar results were reported by VÖLCKER (1914), with a 62% increase in yield of wheat. STUTZER (1914) demonstrated that a lead concentration of 0.5 g/L of nutrient solution induced optimal growth in various crop plants. Using higher lead concentrations, the roots and leaves showed damage phenomena. The yield of sugar was increased when STUTZER (1914) fertilized sugarbeet fields with four kg of lead nitrate/A. On the other hand, potato plants were very sensitive towards lead. Tuber mass as well as starch content were lowered.

EHRENBERG and NOLTE (1916) demonstrated that the nutrient solutions used by STUTZER in his growth experiments contained sufficient PO_4^{3-} and SO_3^{2-} to precipitate all the lead. According to BERRY (1923), the positive effect on plant growth was entirely due to the nitrate ion. Under natural conditions in the field even higher concentrations of lead were harmless because a great part of the lead was retained by anionic groups of the soil (GILBERT 1926).

SCHARRER and SCHROPP (1936) demonstrated that the sensitivity to lead of various cultured grasses varies greatly. Among the plant species tested, *Zea mays* (maize) and *Secale cereale* (rye) were not sensitive to lead acetate concentrations of 10^{-10} to 100 mg of Pb^{2+}/L). Barley, oats, and especially wheat, however, showed damage symptoms even at low lead concentrations. A difference in lead sensitivity of *Hordeum vulgare* (barley) and *Phaseolus vulgaris* (bean) was demonstrated by SUCHODOLLER (1967). A concentration of $10^{-1}M$ $Pb(NO_3)_2$ initiated in bean plants visible damage as soon as 30 min after application. Due to the loss of turgor pressure, the leaves were folded downwards. While the bean plants were dead after 24 h, this occurred in barley plants only after four days. In the presence of a lead concentration of $10^{-5}M$ an enhanced growth activity, as compared to untreated control plants, was observed. The limiting concentration which caused neither damage effect nor growth stimulation was $10^{-3}M$. Interestingly, the leaves inserted at the basal part of the stem showed no response to high lead concentrations, whereas all those leaves which were not fully differentiated at the beginning of the experiments showed chlorosis. At all lead concentrations tested, a decrease in dry matter production of the above-ground plant parts was measured. The weight of the roots, however, was increased when low lead concentrations were present.

The effect of lead ions on the process of seed germination was investigated by DILLING (1926) (see also BELL 1924). Lead (as nitrate and acetate) in concentrations exceeding 0.01% reduced the ability of seeds of cress and mustard to germinate. Concentrations ranging from 0.1 to 0.2% completely inhibited seed germination. The ability to germinate was maintained over a period of 18 days. Upon tranfering the seeds into lead-free medium, the germination was restored.

The influence of lead on the root system is highly pronounced. Application of higher lead concentrations inhibits root growth to a great extent. BONNET (1922) observed on cereals the occurrence of remarkably thick roots with abnormally long root hairs after lead treatment. Studies on hyacinth bulbs (BELL 1924) showed that even very low lead concentrations reduced root growth. Other authors (see GRIFFITH 1919, RASMUSSEN and HENRY 1963) reported an abnormal thinning of the roots, accompanied by a reduction of root hair number.

The influence of different lead concentrations on the root ontogeny of germinating lettuce seeds is depicted in Figure 6. A 93% inhibition

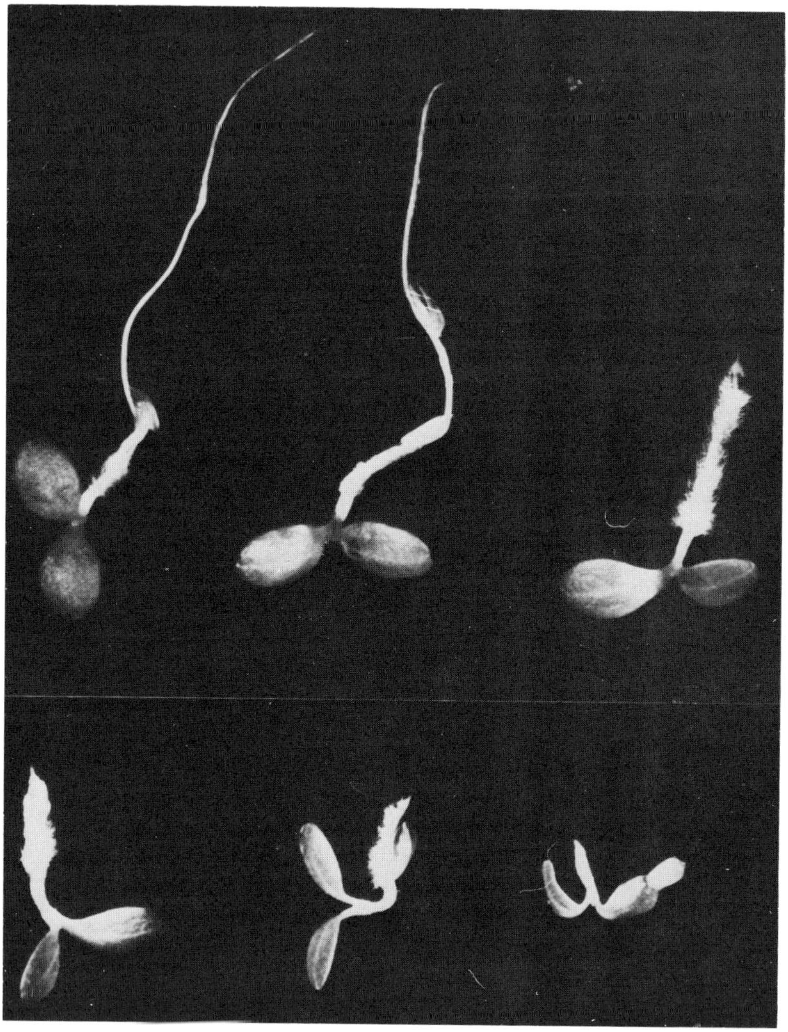

Fig. 6. Reduction of root development of lettuce seedlings (96-h old) by different concentration of lead nitrate from left to right: top = 0, 50, and 100 and bottom = 250, 500, and 1,000 μM $Pb(NO_3)_2$ (from HAMPP 1973 b).

was observed at a concentration of one m*M*, as compared with untreated plants. Possibly the lead reduces the rate of mitosis of the root meristem cells. HAMMETT (1928 d and e) showed that tissues with a low mitose activity contained only little lead in their nuclei. On the other hand, actively dividing nuclei exhibited a high affinity towards lead which is probably due to a high nucleotide level. In *Allium cepa* (onion) and *Zea mays* (maize) an inhibition of mitosis accompanied by a reduced growth was observed when the nutrition solutes contained rising quantities of lead.

In common, one can state that the sensitivity of differentiating young plant tissues towards lead is much higher than that of mature tissues.

The influence of lead ions on the rate of transpiration has been investigated by several authors. According to BONNET (1922), plants reduced their transpiration in the presence of lead ions (see also RASMUSSEN and HENRY 1963). Using spinach leaves, HAMPP (1973 b) demonstrated a pronounced inhibition of transpiration and water uptake in the presence of very low lead concentrations (see Fig. 7). A rise in lead concentration (exceeding 100 μM Pb^{2+}) was not accompanied by a proportional increase of inhibition. Carbon dioxide-change measurements using infrared carbon dioxide analyzer techniques also showed a pronounced inhibition of gas turnover when minor quantities (25 μM) of lead were applied. The inhibition of

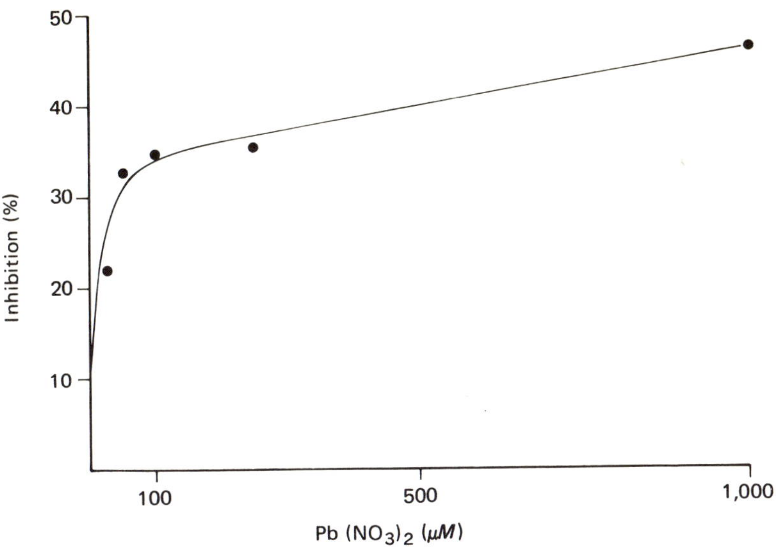

Fig. 7. Percentage of transpiration inhibition in spinach leaves by different concentrations of lead; the incubation period was six h; the inhibition of water uptake exhibits the same pattern (from HAMPP 1973 b).

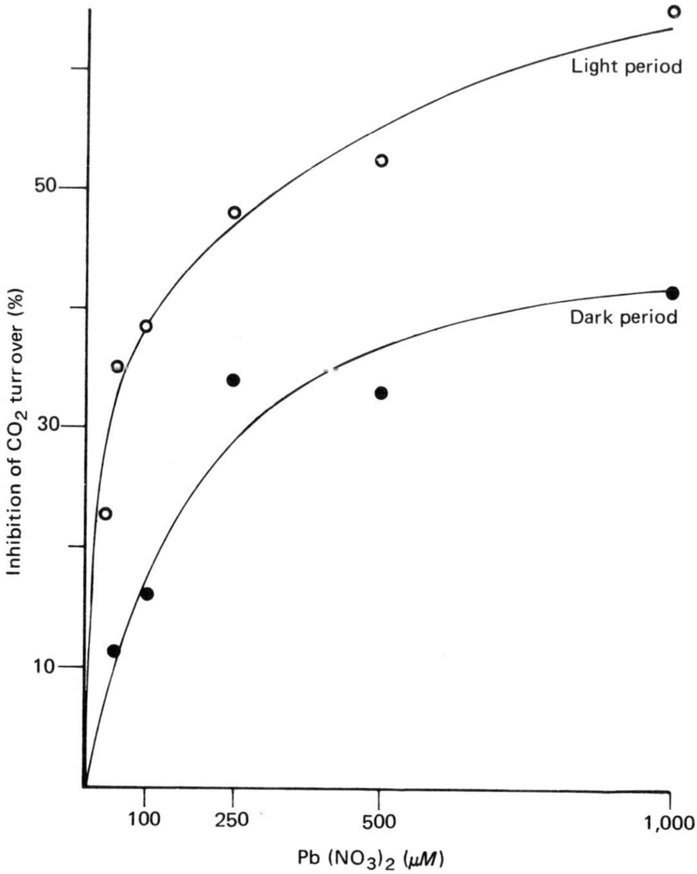

Fig. 8. The inhibition of CO_2-uptake (light period) and CO_2-release (dark period) of spinach leaves in the presence of different concentrations of lead expressed in % *versus* uncontaminated controls (from HAMPP 1973 b).

CO_2-uptake in the light period was higher than the CO_2-release in the dark period (see Fig. 8).

From these findings one can conclude that the process of photosynthesis is more sensitive to lead than the respiration apparatus. In this connection, the fact that stomatal cells accumulate lead is surely of physiological significance. HOOPER (1937) observed that leaves of *Phaseolus* plants sprayed with a lead sulfate solution closed their stomata. As the movement of stomata cells is accompanied by changes in permeability, one could conclude that lead ions alter biomembranes. According to this, lead in the form of DL-cysteine-chelate initiated the leakage of UV-absorbing substances from *Aspergillus niger* cells which were different from those of lead-free cultures (ZLOCHEVSKAYA 1968). One of these substances was identified as uracil. In barley seedlings the inhibitory effect of adenine on the potassium accumulation of the roots was enhanced by $PbCl_2$ (VAKHMISTROV 1969). Measurements of efflux on *Beta vulgaris* (red beet) slices (HAMPP 1973 b) demonstrated that lead concentrations up to 500 μM (in the form of lead nitrate) inhibited exchange of betalaine. With increasing lead concentrations a linear decrease of the inhibitory effect was observed. At a concentration of one mM the efflux rate was significantly higher than in lead-free controls. The result presented till now indicate that lead probably changes the permeability of cell membranes.

VI. Biochemistry

The action of lead on plant enzymes is a neglected aspect. KOEPPE and MILLER (1970) investigated the effect of lead on substrate oxidation in mitochondria isolated from three-days-old etiolated corn shoots. Lead chloride (ten to 62 mM) stimulated oxidation of exogenous NADH by 174 to 640%, whereas lead chloride at 12.5 mM inhibited succinate oxidation by more than 80%. When inorganic phosphate was included in reaction mediums, the subsequent addition of lead was without effect due to the low solubility of lead phosphate. If addition of lead was followed by addition of phosphate the inhibition of succinate oxidation by lead was released, but there was no reduction in the stimulation of oxidation of NADH by lead. From these data the authors concluded that plants which contain high concentrations of lead are unlikely to be affected by these concentrations due to the precipitation of the lead if sufficient phosphate is present. It is significant that the effect of lead was observed at very low concentrations and that there was a sharp increase in the magnitude of influence within a small range of concentrations.

Photosynthesis, a process characteristic for green plants, is also influenced by lead ions. The intensity of $^{14}CO_2$-fixation by isolated spinach chloroplasts was reduced by all lead concentrations tested, from 2 to 200 μM $Pb(NO_3)_2$ (HAMPP *et al.* 1973 a). The inhibition was noncompetitive in nature. Various ^{14}C-labeled compounds (phosphoglyceric acid, malic acid, citric acid, aspartic acid, and sucrose) were lowered to the same amount in comparison with untreated controls. The concentrations of sugar diphosphates, however, were lowered to a much higher degree. This enhanced decrease in concentration is possibly due to a lack of free phosphate in the presence of lead ions, which shift the equilibrium balance towards phosphatase-reactions. It was demonstrated that all lead concentrations tested reduced the ATP-synthesis in illuminated chloroplasts.

The influence of lead ions on partial reactions of photosynthesis was investigated in more detail by MILES *et al.* (1972). In contrast to the electron transport of photosystem I (PS I wherein the electron flow from DCIP to methylviologen was measured), photosystem II (PS II, wherein the Hill reaction and the change in chlorophyll fluorescence were measured) was inhibited by lead salts. The measured fluorescence points towards the possibility that the inhibition occurs predominantly at the oxidative part of PS II, localized between primary electron donor and the site of water oxidation. According to MILES *et al.* (1972), the only reasonable interpretation of the fluorescence inhibition is that the effect is on the oxidizing side of the photoact, since maximum fluorescence associated with reduction of the quencher of fluorescence (Q) was not reached. In contrast to MILES *et al.* (1972), HAMPP (1973 b) demonstrated that PS I was inhibited in the presence of lead ions. The inhibition, however, was less pronounced than that of PS II (see Fig. 9).

The influence of lead on the photosynthetic electron flow might be due to the different redox potentials. PS I is able to reduce substances which show a redox potential up to —650 mV (ZWEIG and ARNON 1965). Since PS I is relatively unspecific at the acceptor site, it reduces

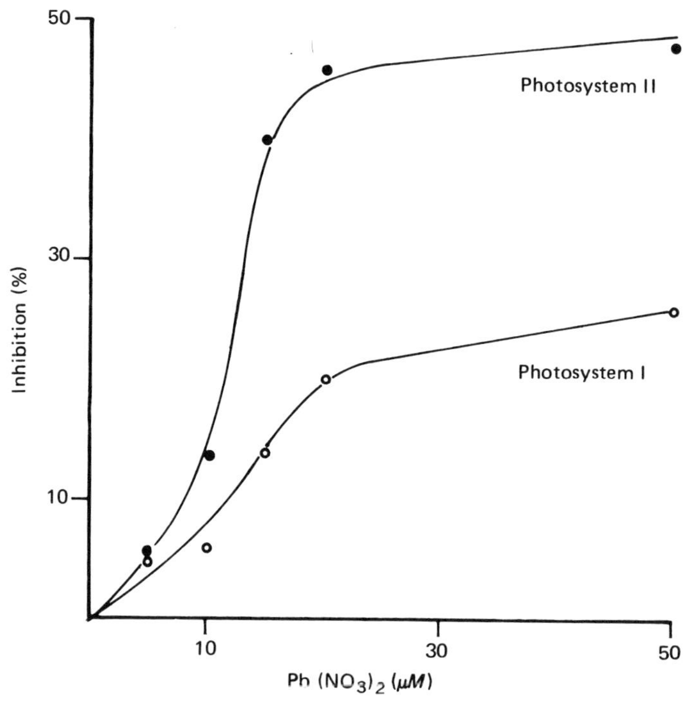

Fig. 9. Inhibitory effect of lead ions (in % of controls without lead) on the potassium ferricyanide reduction of photosystems I and II in isolated spinach chloroplasts (from HAMPP 1973 b).

even unphysiological compounds (see TREBST 1970). Due to more recent findings, the acceptor of PS II shows a redox potential of —150 mV (AMESZ *et al.* 1970). For lead ($Pb \rightleftharpoons 2e + Pb^{2+}$) the redox potential is —126 mV (SEEL 1960). Lead ions are thus reducable by both photosystems.

A more detailed insight into the mechanism of lead action on photosynthesis must come from studies on enzyme activities involved in this process. HAMMETT (1928 d) already mentioned the interaction of lead ions and sulfhydryl groups. Extensive studies on animal enzymes (see VALLEE and ULMER 1972) showed that there is partly a high sensitivity towards lead.

In root and leaf homogenates of spinach the changes in activities in the presence of lead nitrate (5 to 200 μM) of the following enzymes were studied (HAMPP *et al.* 1973 b): ribulosediphosphate carboxylase (EC 4.1.1.39) (see Fig. 10), ribulokinase (EC 2.7.1.16) (see Fig. 11), pyruvate kinase (EC 2.7.1.40) (see Fig. 12), lactate dehydrogenase (EC 1.1.1.27) (see Fig. 13), and glyceraldehydephosphate dehydrogenase (EC 1.2.1.12) (see Fig. 14). Thus is shown that (a) ribulosediphosphate carboxylase was inhibited even at a lead nitrate concentration of five μM (b) lactate dehydrogenase showed a similar but less pronounced decrease in activity, (c) pyruvate kinase activity was enhanced by all lead nitrate concentrations tested with an optimum

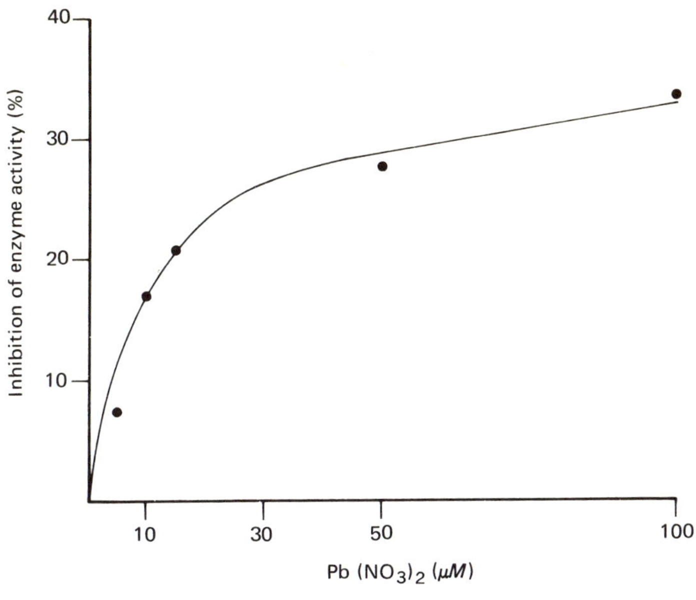

Fig. 10. Inhibition of ribulose-1,5-diphosphate carboxylase activity in sonicated chloroplasts from spinach leaves by different concentrations of lead nitrate (from HAMPP *et al.* 1973 b).

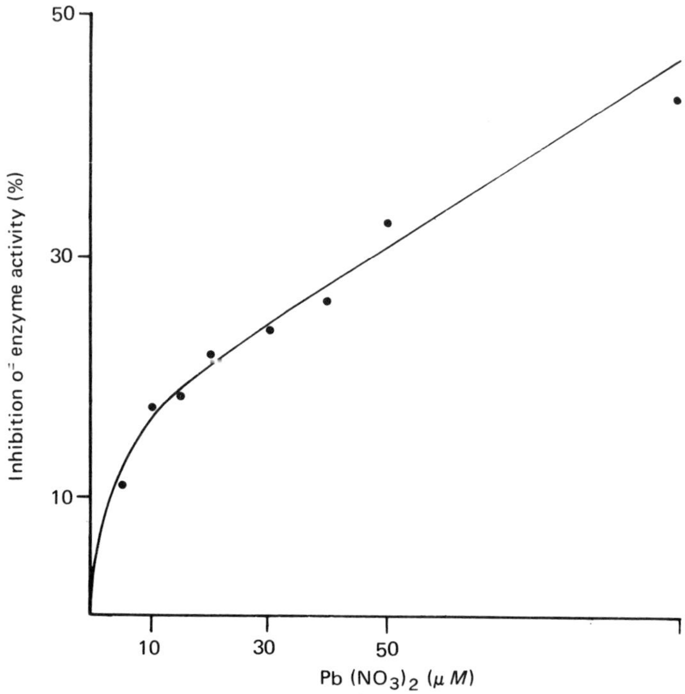

Fig. 11. Inhibition of ribulokinase activity in isolated chloroplasts of spinach leaves by different concentrations of lead nitrate (from HAMPP *et al.* 1973 b).

(more than 300% of the activity of controls) at a $Pb(NO_3)_2$ concentration of 20 m*M*, (d) in the system glyceraldehydephosphate dehydrogenase/glycerokinase there was a superposition of inhibition and activation. Thus, in (d) in the range of medium lead concentrations (100 to 500 μM Pb^{2+} for NAD-dependent glyceraldehydephosphate dehydrogenase, 100 to 1,100 μM Pb^{2+} for NADP-dependent glyceraldehydephosphate dehydrogenase) the activation of the overall reaction dominates; exceeding these concentrations resulted in a pronounced inhibition.

The enzymes mentioned are as lead sensitive as the most sensitive enzymes from animal sources such as lipoamide dehydrogenase, certain ATP-ases, and the δ-aminolaevulinic dehydratase (see VALLEE and ULMER 1972).

The activation of NAD- and NADP-dependent triosephosphate dehydrogenase systems in the presence of lead ions (see above) might be possibly due to the high affinity of the heavy metal to free phosphate groups. During the reduction of phosphoglyceric acid, this acid is first phosphorylated to 1,3-diphosphoglyceric acid and then hy-

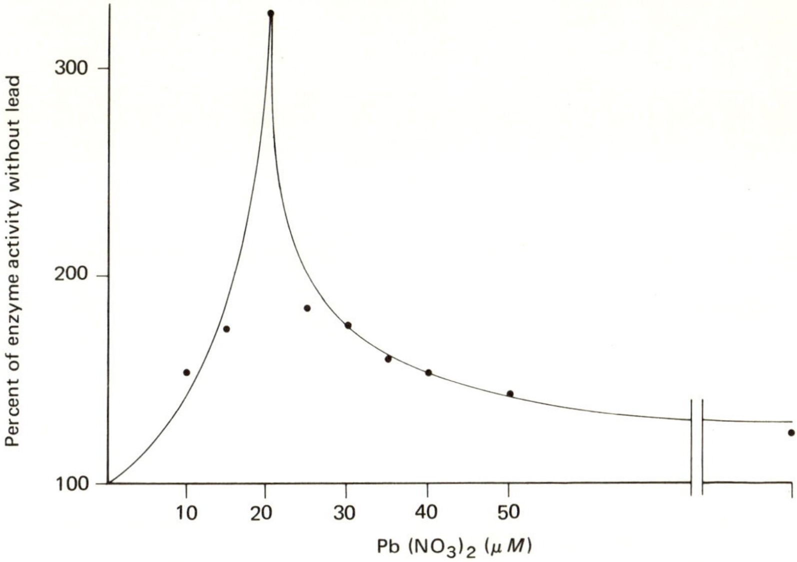

Fig. 12. Influence of different lead concentrations on the reaction sequence phosphoenolpyruvate-pyruvate-lactate (NADH oxidation was taken as measure of enzyme activity) in comparison to assays without lead (from HAMPP *et al.* 1973 b).

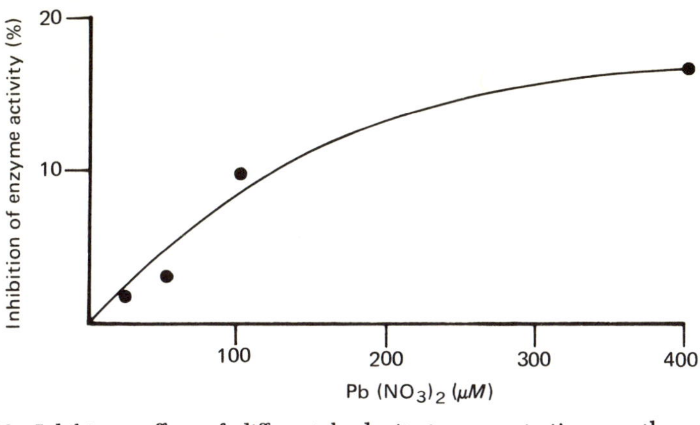

Fig. 13. Inhibitory effect of different lead nitrate concentrations on the activity of lactate dehydrogenase in isolated spinach chloroplasts (from HAMPP *et al.* 1973 b).

drated to glyceraldehyde-3-phosphate with splitting off the phosphate group in position 1. One can speculate that while lead removes the phosphate, a higher turnover of phosphoglyceric acid occurs. This activation is in competition with the inhibition of the enzymes by blocking functional SH-groups. The inhibition is greater than the acti-

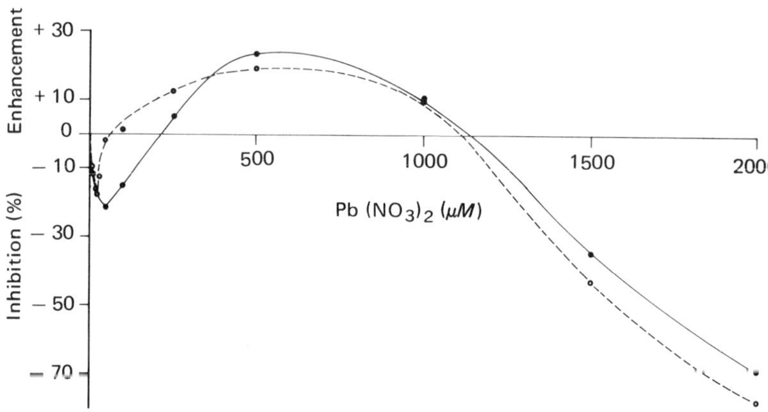

Fig. 14. Changes in the activity of the enzyme system NADP-dependent glyceraldehydephosphate dehydrogenase/phosphoglyceric kinase in isolated spinach chloroplasts in the presence of different lead concentrations: chloroplasts were illuminated (dashed line) or darkened (solid line) before incubation (from HAMPP *et al.* 1973 b).

vation when a critical lead concentration is exceeded. Besides the reaction with SH-groups which runs parallel with a change of conformation of the protein, the blocking of carboxyl groups by lead ions seems to play a major role in metabolism (EDELHOCH 1958, AFONOVA 1958). Under the influence of lead ions, the concentrations of a number of metabolites such as acetyl-coenzyme A (BERSIN 1958), malic acid and other soluble organic acids (RASMUSSEN and HENRY 1963, see also ZLOCHEVSKAYA 1968), and kardenolides (PONERT 1964) in plant tissue were lowered.

Summary

With the beginning of the Industrial Revolution increasing quantities of lead emerged into the environment. At present approximately 98% of the lead in the ecosystem derives from the combustion of motor fuels. In the form of aerosols suspended for a long time in the air (50 to 90% of the lead particles are less than one μm in diameter), the lead is distributed worldwide with the aid of winds.

The lead concentration in plants depends on environmental conditions. In the vicinity of roads with high traffic load or in lead-processing industries, the lead content of plant tissues is often extraordinarily high. With increasing distance from the lead emission source, the lead content in plants decreases. In general, plants which grow leeward to a lead source exhibit higher concentrations of the heavy metal than those growing on the windward side. The amounts of lead deposited on the surface of plants (removable by water wash) differ within a wide range, for the structure of the surface of plants is of importance. Plant organs which possess a hairy or a rough surface accumulate significantly more lead than those with a smooth surface. Epiphytic plants often contain abnormaly high quantities of lead which derive from the atmosphere. Plant organs which show an intensive gas exchange with the surrounding atmosphere contain more lead than storage tissue. For the nutrition of man, it is important that protected portions of plants, which in most cases are edible, contain only low lead concentrations, which show no correlation to the intensity of lead pollution.

During the vegetative period there is a fluctuation in lead content. During the enhanced growing period in spring, the relation between lead quantity/dry weight decreases. With progressive differentiation of the plant tissue, this quotient rises; a maximum is reached in winter time.

Within the cell, lead seems to be bound to polyuronic acids of the cell wall. Organelles as mitochondria, chloroplasts, and also nuclei often show high concentrations of lead. It is likely that lead is bound to phosphate groups.

Within a plant species, there are lead-resistant and lead-sensitive breeds. The resistant representatives are able to exist in soils containing up to one % of lead. This resistance seems to be genetically fixed. Lead salts administered to the plant via the root system are translocated into the above-ground parts only in small amounts. The main part of the applied metal is precipitated in the root. The amount of lead transported from the roots to the stem and leaves depends on the plant species. The uptake of lead by the roots depends on the quality and the pH value of the soil; pH values below 7 cause a higher uptake than those of the alkaline range.

A translocation within the plant body is predominantely apoplastic in nature. Indirect evidence points towards a transport via sieve tubes. In this regard, lead chelate complexes seem to play an important role. The sensitivity towards lead depends on the age of the plant organ: growing leaves are much more sensitive than those already differentiated. Lead damages manifest themselves in an early stage in a loss of turgor pressure, which is correlated to permeability variations of the plasma membrane. In the presence of lead ions, roots react with degeneration of the tissue. It is likely that this degeneration can be explained with a decrease of cell divisions in the root meristems.

The process of photosynthesis is influenced by lead ions. The CO_2-fixation of isolated chloroplasts is reduced. The electron transport of photosystem II is most likely inhibited between the site of primary electron donor and water oxidation. The inhibition of photosystem I is considerably smaller than that of photosystem II. The inhibitory effect of lead ions on the activity of enzymes is due to the blocking of sulfhydryl groups of the proteins and probably due to a change of phosphate levels in the living cells.

References

Afonova, V. N.: Changes in some reactive groups of serum proteins in chronic lead poisoning. Farmakol. Toksikol. **21**, 64 (1958).

Amesz, J. P. van den Bos, and M. P. Dirks: Oxidation-reduction potentials of photosynthetic intermediates. Biochim. Biophys. Acta **197**, 324 (1970).

Atkins, P.: Lead in a suburban environment. J. Air Pollution Control Assoc. **19**, 591 (1969).

Baumhardt, G. R.: Absorption by and growth of corn (*Zea mays* L.) with soil-applied lead. Master's thesis, Univ. Ill. 51 pp. (1970).

Bell, B. W.: Influence of lead on normal and abnormal cell-growth. Lancet **206**, 267 (1924).

Berry, R. A.: The manurial properties of lead nitrate. J. Agr. Sci. **14**, 58 (1923).

Bersin, Th.: Enzymgifte der Autoabgase. Vitalstoffe 3, 160 (1958).

Blanchard, R. L.: Relationship between polonium-210 and lead-210 in man and his environment. In: Proceedings Internat. Symp. "Radioecological Concentration Process," pp. 281–294. New York: Pergamon Press (1966).

Bonnet, E.: Action of soluble lead salts on plants. Compt. Rend. Acad. Sci. (Paris) **174**, 488 (1922).

Bovay, E.: Les dépots de plombs sur la végétation le long des autoroutes. Essai d'affouragement de vaches laitières avec du foin souillé par du plomb. Mitt. aus dem Gebiet der Lebensmitteluntersuchung und Hygiene, Bern **61**, 303 (1970).

Bradshaw, A. D.: Populations of *Agrostis tenuis* resistant to lead and zinc poisoning. Nature **169**, 1098 (1952).

Briggs, D.: Population differentiation in *Marchantia polymorpha* L. in various lead pollution levels. Nature **238**, 166 (1972).

Broyer, T., C. Johnson, and R. Paull: Some aspects of lead in plant nutrition. Plant and Soil **36**, 301 (1972).

Büttner, H.: Erste Ergebnisse von Bleiuntersuchungen in der Luft im Rahmen eine DFG-Meßprogramms. In: Komm. f. Umweltgefahren des Bundesgesundheitsamtes, Arbeitsgruppe Blei, Sitzung v. 2.12.1970; als Manuskript gedruckt v. Verein f. Wasser-, Boden- und Lufthygiene E. V. Berlin: BRD (1972).

Cannon, H. L., and J. M. Bowles: Contamination of vegetation by tetraethyl lead. Science **137**, 765 (1962).

Chow: T. J.: Lead accumulation in roadside soil and grass. Nature **225**, 295 (1970).

——, and J. L. Earl: Lead aerosols in the atmosphere: Increasing concentrations. Science **169**, 577 (1970).

——, and C. C. Patterson: The occurrence and significance of lead isotopes in pelagic sediments. Geochim. Cosmochim. Acta **26**, 263 (1962).

Clymo, R. S.: Ion exchange in Sphagnum and its relation to bog ecology. Ann. Bot. **27**, 309 (1963).

Committee on Biologic Effects of Atmospheric Pollutants: Lead. Airborne lead in perspective. Washington, D.C.: Nat. Acad. Sci. (1972).

Crafts, A. S., and C. E. Crisp: Phloem transport in plants, p. 191, San Francisco: Freeman & Co. (1971).

Daines, R., H. Motto, and D. Chilko: Atmospheric lead, its relationship to traffic volume and proximity to highways. Environ. Sci. Technol. **4**, 318 (1970).

Dale, A. G.: A study of aging of Pb aerosols. II. A numerical model simulating coagulation and sedimentation of a leaded aerosol in the presence of an unleaded background aerosol. Atmos. Environ. **6**, 451 (1972).

——, and J. W. WINCHESTER: A study of aging of Pb aerosols. I. Observation. Atmos. Environ. **6**, 443 (1972).

DANIELSON, L.: Gasoline containing lead. Ecolog. Res. Committee, No. 6. Stockholm: Nat. Sci. Res. Council, (1970).

DILLING, W. I.: Influence of lead and the metallic ions of copper, zinc, thorium, beryllium and thallium on the germination of seeds. Ann. Applied Biol. **13**, 160 (1926).

DJURIČ, D., Z. KERIN, L. GRAOVAC-LEPOSAVIC, L. NOVAK, and M. KOP: Environmental contamination by lead from a mine and smelter. A preliminary report. Arch. Environ. Health **23**, 275 (1971).

DUNN, J. T., and H. C. L. BLOXAM: The occurrence of lead, copper, zinc, and arsenic compounds in atmospheric dust and the source of these impurities. J. Soc. Chem. Ind. **52**, 189 (1933).

EDELHOCH, H.: The effect of various protein denaturants on the kinetics of pepsin inactivation. J. Amer. Chem. Soc. **80**, 6648 (1958).

EHRENBERG, P., and O. NOLTE: Kritische Gedanken über die Empfehlung von Reizdüngemitteln. Fühlings Landw. Ztg. **65**, 559 (1916).

Eidg. Kommission für Lufthygiene: Das Problem der Benzinverbleiung. Beilage B, Nr. 3/1971 zum Bulletin des Eidg. Gesundheitsamtes, Bern (1971).

EVERETT, J. C., C. L. DAY, and D. REYNOLDS: Comparative survey of lead at selected sites in the British Isles in relation to air pollution. Food Cosmetic Toxicol. **5**, 29 (1967).

FIDORA, B.: Der Bleigehalt von Pflanzen verkehrsnaher Standorte in Abhängigkeit von der Vegetationsperiode. Ber. Deutsch. Bot. Ges. **85**, 219 (1972).

FLETCHER, K.: Direct determination of lead in plant materials by atomic absorption spectrophotometry. J. Sci. Food Agr. **22**, 260 (1971).

FRANCIS, C. W., G. CHESTERS, and L. A. HASKIN: Determinations of 210-Pb mean residence time in atmosphere. Environ. Sci. Technol. **4**, 586 (1970).

GARBER, K.: Luftverunreinigung durch schwermetallhaltige Stäube. Wirkung auf Pflanzen. Landw. Forschung **25**, 59 (1970).

—— Luftverunreinigungen, eine Literaturübersicht. In W. Bosshard (ed.): Ber. Nr. 102 der Eidg. Anstalt für das forstliche Versuchswesen, CH-8903 Birmensdorf (1973).

GILBERT, P.: Mineral nutrition and the balance of life. Ann. Applied Biol. **13**, 246 (1926).

GOREN-SUCHODOLLER, A., and H. WANNER: Die Absorption von Blei und Mangan durch Wurzeln von Hordeum vulgare. Z. Pflanzenphysiol. **61**, 122 (1969).

GRIFFITH, J. J.: Influence of mines upon land and lifestock in Cardiganshire. J. Agr. Sci. **9**, 366 (1919).

HAMMETT, F. S.: Studies in the biology of metals. I. The localisation of lead by growing roots. Protoplasma **4**, 183 (1928 a).

—— Studies in the biology of metals. III. The localization of lead within the cell of the growing root. Protoplasma **5**, 135 (1928 b).

—— The influence of lead on mitosis and cell size in the growing root. Protoplasma **5**, 535 (1928 c).

—— The nature of lead compound deposited in the growing root. Protoplasma **5**, 447 (1928 d).

HAMPP, R.: Bleigehalt von Blattspreiten (*Acer platanoides* L.) als Indikator für die verkehrsabhängige Bleibelastung im Stadtgebiet von München. Ber. Bayer. Bot. Ges. **44**, 211 (1972 a).

—— Über die Konzentration von Bleiverbindungen in Pflanzenteilen und die Wirkung des Bleiions auf pflanzenphysiologische und biochemische Prozesse. Ph.D. Thesis, München (1973 b).

——, and W. HÖLL: Radial and axial gradients of lead concentration in bark and xylem of hardwoods. Arch. Environ. Contam. Toxicol. In press (1974).

——, H. ZIEGLER, and I. ZIEGLER: Die Wirkung von Bleiionen auf die $^{14}CO_2$-Fixietrung und die ATP-Bildung von Spinatchloroplasten. BBP **164**, 126 (1973 a).

—— —— —— Der Einfluβ von Bleiionen auf Enzyme des reduktiven Pentosephosphatcyclus. BPP **164**, 588 (1973 b).

HAMMOND, P. B., and A. L. ARONSON: Lead poisoning in cattle and horses in the vicinity of a smelter. Ann. N.Y. Acad. Sci. **111**, 595 (1964).

HASSLER, A.: Retention of copper in soil. Mitt. Lebensm. Hyg. (Bern) **34**, 79 (1943).

HEILENZ, S.: Untersuchungen über den Bleigehalt von Pflanzen an verkehrsreichen Standorten. Landw. Forschung **25**, 73 (1970).

HEVESY, G.: LIII. The absorption and translocation of lead by plants. A contribution to the application of the method of radioactive indicators in the investigation of the change of substance in plants. Biochem. J. **17**, 439 (1923).

HIRSCHLER, D. A., and L. F. GILBERT: Nature of lead in automobile exhaust gas. Amer. Med. Assoc. Environ. Health, p. 297 (1964).

HOOPER, M. C.: An investigation of the effect of lead on plants. Ann. Applied Biol. **24**, 690 (1937).

HOPKINSON, J. M., R. H. WILSON, and B. N. SMITH: Lead levels in plants. Naturwissenschaften **59**, 421 (1972).

JONES, L. H. P., S. C. JARVIS, and D. W. COWLING: Lead uptake from soils by perennial ryegrass and its relation to the supply of essential element (sulphur). Plant and Soil **38**, 605 (1973).

JOWETT, D.: Populations of *Agrostis* ssp. tolerant of heavy metals. Nature **182**, 816 (1958).

—— Population studies on lead tolerant *Agrostis tenuis*. Evolution **18**, 70 (1964).

KEATON, C. M.: The influence of lead compounds on the growth of barley. Soil Sci. **43**, 401 (1937).

KELLER, Th., and H. PREIS: Der Bleigehalt von Fichtennadeln als Indikator einer verkehrsbedingten Luftverunreinigung. Schweiz. Z. Forstwes. **118**, 143 (1967).

——, and R. ZUBER: Über die Bleiaufnahme und Bleiverteilung in jungen Fichten. Forstwiss. Centralbl. **89**, 20 (1970).

KERIN, D.: The determination of heavy metals on plant material for the detection and determination of air pollution damage from metallurgy. Presented XVth Congress Internat. Union Forest Research Org., IUFRO-Working Group on Fume Damage, Sect. 24, Gainsville, Fla. (1971).

——, and Z. KERIN: Bleikontamination von Milch und Honig durch Bleiaerosole der Industrie. Protectio vitae (Stuttgart, Kohlhammer) **82**, 61 (1971).

KERIN, Z.: Verunreinigungen von Gemüse aus Emissionen einer Bleihütte. Qual. Plant. Mat. veget., Den Haag **15**, 372 (1968).

—— Relationship between the lead concentration in the tubers and the soil (Slovakian, summary in English). Agrochemija, Beograd, No. 9/10, p. 341 (1970).

KLOKE, A., and H. O. LEH: Verunreinigungen von Kulturpflanzen mit Blei aus Kraftfahrzeugabgasen. In: Air pollution. Proc. 1st Eur. Congress on Air Pollution on Plants and Animals. Wageningen, 1968, p. 295 (1969).

——, and K. RIEBARTSCH: Verunreinigungen von Kulturpflanzen mit Blei aus Kraftfahrzeugabgasen. Naturwissenschaften **51**, 369 (1964).

KOEPPE, D. E., and R. J. MILLER: Lead effects on corn mitochondrial respiration. Science **167**, 1376 (1970).

LAGERWERFF, J. V., and A. W. SPECHT: Contamination of roadside soil and vegetation with cadmium, nickel, lead, and zinc. Environ. Sci. Technol. **4**, 583 (1970).

—— —— Uptake of cadium, lead and zinc by radish from soil and air. Soil Sci. **111**, 129 (1971).

Laveskog, A.: A method for determination of tetramethyl lead (TML) and tetraethyl lead (TEL) in air. Presented 2nd Internat. Clean Air Congress, Internat. Union Air Pollution Prevention Assoc., Washington, D.C. (1970).

Lee, J. A.: Lead pollution from a factory manufacturing antiknock compounds. Nature **238**, 165 (1972).

Leh, H. O.: Verunreinigungen von Kulturpflanzen mit Blei aus Kraftfahrzeugabgasen. Gesunde Pflanzen **18**, 21 (1966).

Lipman, C. B., and P. S. Burgess: The effect of Cu, Zn, Fe, and Pb salts on ammonification and nitrification in soils. Univ. Calif. Publ. Agr. Sci. **1**, 127 (1914).

Maclean, A. J., R. L. Halstead, and B. J. Finn: Extractability of added lead in soils and its concentrations in plants. Can. J. Soil Sci. **49**, 327 (1969).

Martinez, J. D., M. Nathany, and V. Dharmarajan: Spanish moss, a sensor for lead. Nature **233**, 564 (1971).

Miles, C. D., J. R. Brandle, D. J. Daniel, O. Chu-Der P. D. Schnare, and D. J. Uhlik: Inhibition of photosystem II in isolated chloroplasts by lead. Plant Physiol. **49**, 820 (1972).

Mitchell, R. L., and J. W. S. Reith: The lead content of pasture herbage J. Sci. Food. Agr. **17**, 437 (1966).

Motto, H. L., R. H. Daines, D. M. Chilko, and C. K. Motto: Lead in soils and plants: Its relationship to traffic volume and proximity to highways. Environ. Sci. Technol. **4**, 231 (1970).

Mueller, P. K.: Discussion: Characterization of particulate lead in vehicle exhaust-experimental techniques. Environ. Sci. Technol. **4**, 248 (1970).

Murozumi, M., T. J. Chow, and C. C. Patterson: Chemical concentrations of pollutant lead aerosols, terrestrial dusts, and sea salts in Greenland and Antarctic snow strata. Geochim. Cosmochim. Acta **33**, 1247 (1969).

Nilsson, J.: Accumulation of metals in spruce needles and needle litter. Oikos **23**, 132 (1972).

Page, A., and T. J. Ganje: Accumulation of lead in soils for regions of high and low motor vehicle traffic density. Environ. Sci. Technol. **4**, 140 (1970).

—— ——, and M. S. Joshi: Lead quantities in plants, soils, and air near some major highways in southern California. Hilgardia **41**, 1 (1971).

Paluch, J., and St. Karweta: Die Akkumulierung von Zink und Blei im Boden und in Pflanzen. Referate VI. Internat. Arbeitstg. Forstl. Rauchschaden-Sachverst., Katowiče 9. - 14. 9. 1968, p. 127 (1968).

—— —— Zink- und Bleiimmissionen aus einem Hüttenkombinat und ihre Wirkung auf Pflanzen und Böden. Ochrana powietrza **6**, 2 (1970).

Patterson, C. C.: Contaminated and natural lead environments of man. Arch. Environ. Health **11**, 344 (1965).

Peterson, P. J.: The distribution of zinc-65 in *Agrostis tenuis* Sibth. and *A. stolonifera* L. tissues. J. Exp. Bot. **20**, 863 (1969).

Prát, S.: The absorption of lead by plants. Amer. J. Bot. **14**, 633 (1927).

Ponert, J.: Einfluβ von Blei, Silber, Kupfer und Kalium auf den Gehalt an Kardenoliden in Blättern. Naturwissenschaften **51**, 320 (1964).

Quinche, J., R. Zuber, and E. Bovay: Les dépots de plomb provenant des gaz d'échappement des véhicules automobiles le long des routes àforte densité de circulation. Phytopathol. Z. **66**, 259 (1969).

Raines, D. W.: Lead accumulation by wild oats (*Avena fatua*) in a contaminated area. Nature **233**, 210 (1971).

Rasmussen, E. K., and W. H. Henry: Effects of lead on the growth of sweet orange seedlings in nutrient solution cultures. Proc. Soil Sci. Soc. (Fla.) **23**, 70 (1963).

Report of the *Eidgenössische Komission für Lufthygiene* (1971).

Report of the *Committee on Biological Effects of Atmosphere Pollutant* (1972).

ROBINSON, E., F. L. LUDWIG, J. E. DEVRIES, and P. E. HOPKINS: Variations of atmospheric lead concentrations and type with particle size. Final Report PA-4211, Stanford Res. Inst. (1963).

ROHMEDER, E., and E. WEBER: Fichtenvergiftung durch Flugzeugabgase. Forstwis. Centbl. **89**, 335 (1970).

RÜHLING, A., and G. TYLER: Sorption and retention of heavy metals in the woodland moss *Hylocomium splendens* Hedw. Br. et Sch. Oikos **21**, 92 (1970).

SCHARRER, K., and W. SCHROPP: Über die Wirkung des Bleis auf das Pflanzenwachstum. Z. Pfl. Erneahr. Düng. **43**, 34 (1936).

SEEL, F.: Grundlagen der analytischen Chemie, 2. Aufl. Weinheim: BRD (1960).

SINGER, M. J., and L. HANSON: Lead accumulation in soils near highways in the Twin Cities metropolitan area. Soil Sci. Soc. Amer. Proc. **33**, 152 (1969).

SKAAR, H., E. OPHUS, and B. M. GULLVÅG: Lead accumulation within nuclei of moss leaf cells. Nature **241**, 215 (1973).

SOMMER, G., A. ROSOPULO, and A. KLEE: Die Bleikontamination von Pflanzen und Böden durch Kraftfahrzeugabgase. Z. Pflanzenernährung, Düngung u. Bodenkunde **130**, 193 (1971).

STEWART, J.: Chelation in the absorption and translocation of mineral elements. Ann. Rev. Plant Physiol. **14**, 295 (1963).

STOKLASA, J. M.: Influence of uranium and lead upon vegetation. Compt. rend. Acad. Sci. (Paris) **156**, 153 (1913).

STUTZER, A.: Weitere Erfahrungen mit der Anwendung sog. Reizstoffe. Dtsch. Landw. Pr. **41**, 1 (1914).

SUCHODOLLER, A.: Untersuchungen über den Bleigehalt von Pflanzen in der Nähe von Straßen und über die Aufnahme und Translokation von Blei durch Pflanzen. Ber. Schweiz. Bot. Ges. **77**, 266 (1967).

SWAINE, D. J., and R. L. MITCHELL: Trace-element distribution in soil profiles. J. Soil Sci. **11**, 347 (1960).

TAMMES, P. M. L., and J. VAN DIE: Studies on phloem exudation from *Yucca flaccida* Haw. Translocation of macro- and micro-nutrients by the phloem sap stream. Koninkl. Ned. Acad. Wetenschap. Proc. (C) **69**, 656 (1966).

TER HAAR, G. L.: Air as a source of lead in edible crops. Environ. Sci. Technol. **4**, 226 (1970).

——, R. B. HOLTZMAN, and H. F. LUCAS, Jr.: Lead and lead 210 in rainwater. Nature **216**, 353 (1967).

——, and M. A. BAYARD: Composition of airborne lead particles. Nature **232**, 553 (1971).

TREBST, A.: Neue Reaktionen und Substanzen im photosynthetischen Elektronentransport. Ber. Dtsch. Bot. Ges. **83**, 373 (1970).

TSO, T. C., and I. FISENNE: Translocation and distribution of lead-210 and polonium-210 supplied to tobacco plants. Radiation Bot. **8**, 457 (1968).

——, N. HARLEY, and L. T. ALEXANDER: Source of lead-210 and polonium-210 in tobacco. Science **153**, 880 (1966).

TUNNEY, H., G. A. FLEMING, A. N. O'SULLIVAN, and J. P. MOLLOY: Effects of lead-mine concentrates on lead content of ryegrass and pasture herbage. Irish J. Agr. Res. **11**, 85 (1972).

TUOMINEN, Y.: Studies on the strontium uptake on the *Cladonia alpestris thallus*. Ann. Bot. Fenn. **4**, 1 (1967).

TURNER, R. G., and C. MARSHALL: The accumulation of ^{65}Zn by root homogenates of zinc-tolerant and non-tolerant clones of *Agrostis tenuis* Sibth. New Phytol. **70**, 539 (1971 a).

—— —— The accumulation of zinc by subcellular fractions of roots of *Agrostis tenuis* Sibth. in relation to zinc tolerance. New Phytol. **71**, 671 (1971 b).

VALLEE, B. L., and D. D. ULMER: Biochemical effects of mercury, cadmium, and lead. Ann. Rev. Biochem. **41**, 91 (1972).

VAKHMISTROV, D. B.: Combined action of adenine and agents blocking the protein component of cell membranes on potassium accumulation by barley shoots. (In Russian) Fiziol. Rast. **16**, 255 (1969).

VÖLCKER, J. A.: The Woburn pot culture experiments. J. R. Agr. Soc. (Eng.) **75**, 306 (1914).

WARREN, H. V., and R. E. DELAVAULT: Lead in some food crops and trees. J. Sci. Food Agr. **13**, 96 (1962).

WILKINS, D. A.: A technique for the measurement of lead tolerance in plants. Nature **180**, 37 (1957).

YODAIKEN, R. E.: Cited by SKAAR *et al.* (1973).

ZIEGLER, H.: Untersuchungen über die Leitung und Sekretion der Assimilate. Planta **47**, 447 (1956).

ZLOCHEVSKAYA, A.: Toxic effect of a lead complex with DL-cysteine on *Aspergillus niger.* (In Russian). Mikrobiologiya **37**, 848 (1968).

ZUBER, R.: Die Bestimmung des Bleis in pflanzlichem Material mit Hilfe des Atomabsorptionssektrophotometers. Mitt. Lebensmittelunters. Hyg., Bern **63**, 229 (1972).

——, and E. BOVAY: L'assorbimento del piombo da parte della pianta tramite la radice. Esperimenti preliminari sul problema in questione. Parte prima. Schweiz. Landw. Forsch. **11**, 381 (1972).

ZWEIG, G., and M. AVRON: On the oxidation-reduction potentials of the photoproduced reductant of isolated chloroplasts. Biochem. Biophys. Res. Comm. **19**, 397 (1965).

Manuscript received February 13, 1974; accepted March 20, 1974.

Gas chromatographic determination of nitrogen-containing pesticides using the nitrogen flame ionization detector (*N*-FID)

By

H. Maier-Bode* and M. Riedmann**

Contents

I. Introduction

The use of gas chromatographic (glc) methods for purposes of identification and quantitation of pesticides[1] and their residues within the required accuracy and detection limits is directly related to the development of detectors which are selective and sensitive to the chemical structure, elemental constitution, or functional groups.

* Pharmakologisches Institut der Rheinischen Friedrich-Wilhelms-Universität Bonn, 7881 Rickenbach, Tannenweg 7, W. Germany.

** Hewlett-Packard GmbH., 7030 Böblingen, Herrenberger Str. 110, W. Germany.

[1] Common names of pesticides are used throughout the text. Special chemical names are listed in Table XX; examples for trade names are listed in Table XXI.

Table I. *Selectivity to elements and minimum detectability of detectors extensively used in pesticide analysis.*

Detector	Element specificity	Minimum detectability
Electron-capture	Cl, Br, I, P, $-NO_2$, and other groups	pg
Thermionic	P, N	pg
	Cl, Br, I	ng
Microcoulometric	Cl, S, N	ng
Electrolytic conductivity	Cl, S, N	ng
Flame photometric	P, S	ng

Among those detectors widely used in routine glc analysis of pesticides are the electron-capture (EC) detector, the *Dohrmann* microcoulometer, the COULSON conductivity detector, the flame photometric detector, and several different designs of thermionic flame ionization detectors. All of these detectors show selective detection properties. Selective detection facilitates the accomplishment of qualitative and quantitative analysis, particularly at very low sample concentration levels in the presence of concomitant endogenous substances differentiating compounds with similar or the same retention time.

Selective detectors diminish the uncertainty of identification inherent in nonspecific detectors, especially when confirmation of identity is achieved by the use of two liquid phases differing in polarity. A general review of nonselective and selective detectors useful for the detection of pesticides and their residues by glc has been presented by WESTLAKE and GUNTHER (1967).

Table I shows the selectivity to elements and the minimum detectability of those detectors extensively used in pesticide analysis. For the routine glc analysis of nitrogen-containing pesticides the COULSON detector based on the physical principle of electrolytic conductivity (COULSON 1965, 1966 a and b), the *Dohrmann* microcoulometric titration cell based on the coulometric principle described by R. L. MARTIN (1966), and the *Hewlett-Packard* (1969) nitrogen selective flame ionization detector (*N*-FID) based on the thermionic chemical ionization principle are suitable, commercially available, detection devices. The conductivity detector and the coulometric detector involve a pyrolytic degradation and a catalytic reduction of the eluted glc peaks in order to reduce organic nitrogen into ammonia for the purpose of detection. Both detectors have the same detection limit of about one ng of nitrogen and the linear range of about 200 ng. The linearity of the *N*-FID described here ranges from the lowest detection limit of 0.1 to 100 pg of nitrogen/second. The lowest detection limit of 0.1 pg corresponds to a signal height twice the variation in background current.

Since KARMEN and GIUFFRIDA (1964) described a modified flame ionization detector selective for the elements phosphorus and chlorine which was based on the thermionic principle, several different detector designs have been evaluated. An excellent review on thermionic detectors has been presented by BRAZHNIKOV *et al.* (1970). AUE *et al.* (1967) first reported the sensitivity of a thermionic detector to nitrogen in organic molecules. Compared to other designs, the nitrogen flame ionization detector design described here has several advantages in respect to sensitivity, selectivity, stability, linear range, and ease of operation over long periods of time and mechanical roughness. The result is that it is no longer an art, but an exact science making the thermionic detector perform in a satisfactory manner.

II. Detector design, detector operating parameters, and reaction mechanism

a) Detector description

Thermionic detectors differ in the alkali salt source used, in the means of evaporation of the alkali salt, and of collection of the ions formed during the chemical ionization process occurring between the alkai vapour and the heteroelement to be detected. By using, for example, the corresponding alkali halides as the source, the sensitivity for halogens is suppressed and the thermionic detector is rendered insensitive towards halogenated pesticides. The design selected for introducing the alkali vapour into the flame and the type of alkali salt used influences the detector sensitivity and selectivity by several orders of magnitude.

The *Hewlett-Packard* model 15161 B *N*-FID described here operates with rubidium bromide as the alkali source. A rubidium bromide crystal is grown from the molten salt and machined to form a cylinder with a central bore in its longitudinal axis. The crystal fits into the cylindrical collector electrode (see Fig. 1). A mechanical

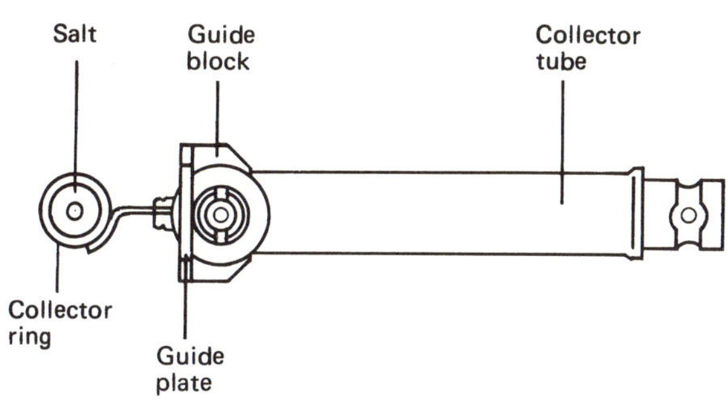

Fig. 1. Collector ring, guide block, and collector tube of the *N*-FID.

means is provided for easy adjustment of the height of the crystal above the flame jet in order to select the desired position of high sensitivity and high selectivity operation. A platinum loop gate-electrode is mounted concentrically with the jet top, enhancing the signal-to-noise ratio by up to a factor of ten when a corresponding negative potential is applied. The gate-electrode potential is adjusted by the zero adjust dial on the selective detector cabinet.

b) Operating mechanism

Nitrogen compounds eluted from the gas chromatograph into the reaction zone located between the flame jet tip and the alkali source surface in contact with the flame are ionized to an extent of about 10^{-1}%, while a hydrocarbon is ionized only to 10^{-4} to 10^{-5}% under operating conditions. These values show that the *N*-FID operates on a principle of signal enhancement for nitrogen-containing compounds and by a reduction in response to other compounds, such as hydrocarbons and oxygenated, sulfurated, and halogenated compounds. Three positions of the crystal with respect to the flame are essential for detector operation in terms of sensitivity and selectivity. The three selected positions are shown in Figure 2 and designated with the numbers 1, 2, and 3. These positions are not mechanically fixed positions, but are areas of the flame through which the collector enters, as it is moved towards the jet tip. By lowering the collector containing the salt crystal into the flame, rubidium bromide evaporates from the surface. The evaporation rate depends on the equilibrium temperature between flame and crystal surface. In a hydrogen/oxygen flame the maximum temperature demonstrated by OHLING *et al.* (1969) was found to be located between about ⅓ and ⅔ of its height. In this position (*i.e.*, position 2) the ionization current maximum is reached since the highest flame temperature corresponds to the highest evaporation rate of the rubid-

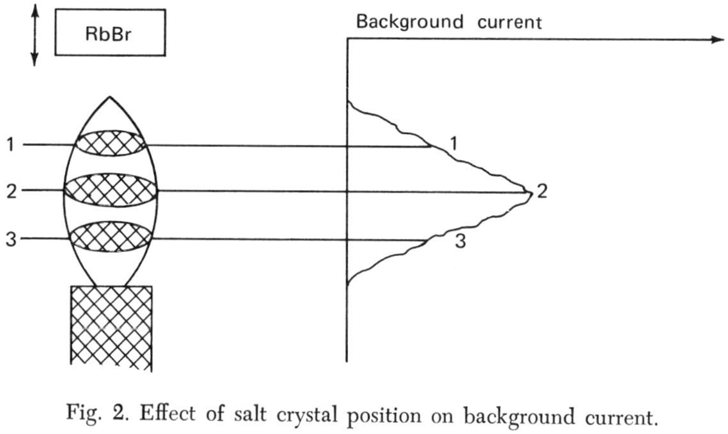

Fig. 2. Effect of salt crystal position on background current.

ium bromide. Above position 1 and below position 3 the flame temperature, and thus the evaporation rate, is much less.

The evaporated rubidium bromide dissociates as follows in the flame:

$$RbBr \overset{\Delta}{\rightleftharpoons} Rb^+ + Br^- \quad (1)$$

The Rb^+ ions form the recorded background current, the ion current being proportional to the evaporation rate of the rubidium bromide. The ion current of the *N*-FID corresponds to about 5 to 8×10^{-9} amp compared to the background current of about 5×10^{-14} amp of the flame ionization detector. The ionization current increases as the collector is lowered from the top position down to position 2 and decreases after further lowering towards the jet, the recorded current corresponding to the flame temperature profile (see Fig. 2).

The ionizaton current is continuously registered by the recorder indicating the salt position in the flame temperature profile. Position 2 corresponds to the recorded maximum ionization and evaporation of the rubidium bromide and represents the position of general operation of the nitrogen detector. In this position sensitivity and selectivity are satisfactory for most practical purposes. In positions 2 and 3 the flame is more quenched by the crystal surface than it is in position 1. This causes the noise to increase when the collector is moved from position 1 into position 2 or 3. Position 1 with the lowest noise level corresponds to the highest detector sensitivity, while position 3 corresponds to the lowest sensitivity because of the large noise level. The signal output (peak height) for nitrogen-containing compounds is approximately constant between positions 1 and 3, while the signal output decreases for compounds not containing a heteroelement to which the detector is sensitive. With the lower collector position 3 the signal output for such compounds decreases up to a factor between 10 and 100. Position 3 therefore corresponds to the highest signal output for nitrogen compounds and the lowest signal output for non-nitrogen compounds (see Fig. 3). The quotient of the signal outputs is defined as the detector selectivity. Table II shows the detec-

Table II. *Element selectivities of the N*-FID.

Elements	Selectivity[a]
N–Cl	1:5,000
N–Br	1:5,000
N–I	1: 200
N–S	1:2,000
N–C	1:5,000
N–P	10:1

[a] See text for definition.

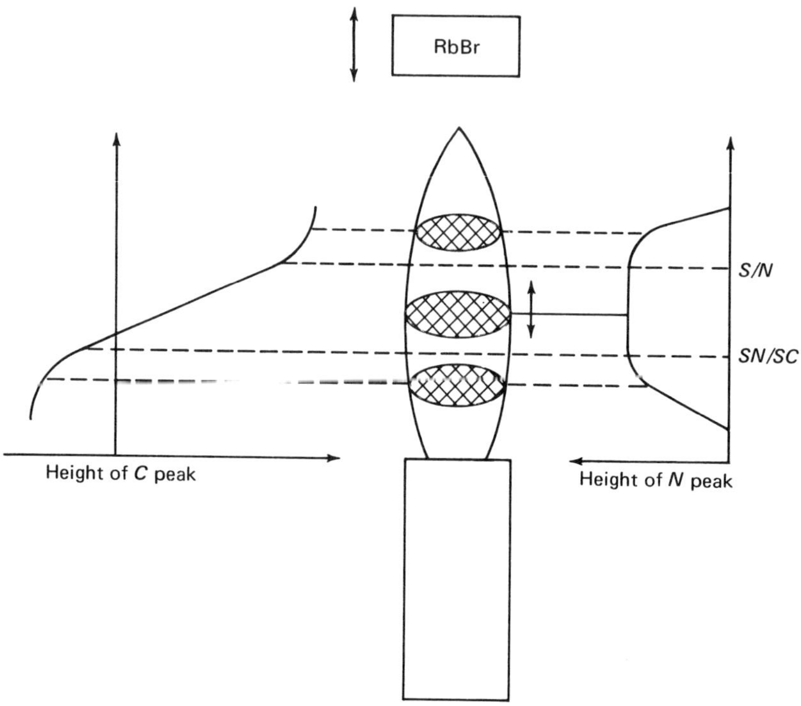

Fig. 3. Effect of salt crystal position on response to carbon and nitrogen.

tor selectivity for different elements, the collector being in position 3. The detector selectivity is defined as the quotient of the signals from equal injected sample quantities, for example:

$$\text{Selectivity} = \frac{1\ \mu\text{g atrazine peak area}}{1\ \mu\text{g octadecane peak area}} = 10^4$$

With the collector in position 3, peak inversion occurs for hydrocarbons or any other compounds for which the detector is not sensitive (Fig. 4). Peak inversion observed in thermionic detectors can easily be understood for the detector design described here, while it is not at all understood for other designs. Peak inversion is not a phenomenon of the thermionic behaviour of the detector. For example, when a hydrocarbon is eluted from the column in an amount of more than about one μg, the flame increases in size and turns back to its initial size after complete elution. During the flame height increase the salt surface is virtually positioned in a lower temperature zone, thus less salt is evaporated causing the background current to decrease. The background current is reestablished when the peak is eluted and the flame is of normal size. The negative peaks show tailing due to the equilibration time required between the flame and the salt surface temperature.

It is not recommended that halogenated compounds be used as solvents for samples that have to be analyzed with the *N*-FID. Halogenated solvents influence the chemical equilibrium described by equation (2) (see later). Bromide ions formed by FID combustion of a brominated solvent shift the equilibrium to the left diminishing the background current formed by the rubidium ion population present under operating conditions. By the presence of chlorine ions formed by flame ionization the quantity of rubidium bromide evaporated from the surface is diminished due to the formation of a rubidium chloride layer on the salt surface. The vapour pressure of the rubidium chloride layer is lower than for rubidium bromide at a given crystal surface temperature. Thus, the recovery time of the *N*-FID background current takes up to several minutes corresponding to the time required for the quantitative evaporation of the rubidium chloride layer formed on the salt surface which inhibits the evaporation of the rubidium bromide. After elution of the halogenated solvent,

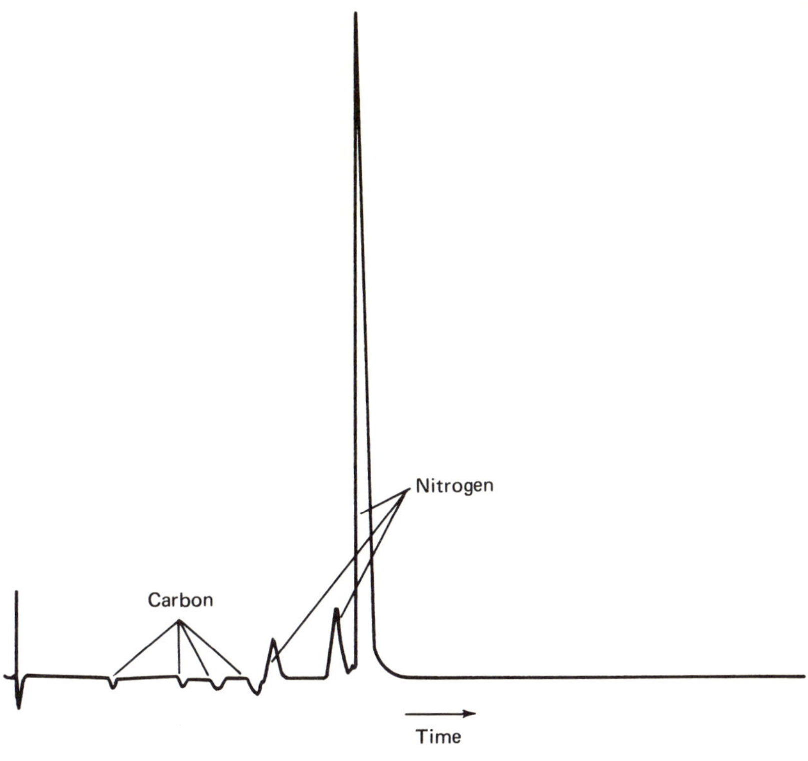

Fig. 4. Chromatogram of carbon and nitrogen sample with crystal in position 3.

causing the negative peak, the baseline recovers slowly to the former background current, and the detector operates with stable sensitivity; however, any pesticide peak eluting during the background current recovery time cannot be evaluated quantitatively.

c) *Reaction mechanism*

One mechanism proposed by KARMEN (1964) asserts formation of active intermediate products which then interact with the salt surface forming volatile compounds of alkali metals which are ionized in the flame. A second mechanism proposed by CREMER (1967) asserts formation of active intermediate products which lead to an increase in the rate of evaporation of the alkali salt which then is ionized to a high extent in the flame. Another mechanism proposed by JANAK *et al.* (1968) for the thermionic reaction is that the addition of phosphorus-containing compounds leads to the formation of active particles in the flame which are capable of reacting with the alkali salt molecule vapour or alkali atoms available in the flame to form atoms of alkali metals which can be readily ionized. JANAK *et al.* also explain the sensitivity of thermionic detectors to elements which, on burning, produce strongly acidic oxides (P_2O_5, As_2O_3, SnO_2) by the fact that the atoms of alkali metals and some other substances interact with the acidic combustion products.

A specific feature of the thermoinic mechanism proposed here is that the basic processes occur in the gas phase between alkali metal atoms and the hetero-organic compound according to the reaction inherent in the LASSAIGNE test. This test is based on the formation of the following ions corresponding to the heteroelement: CN^-, Cl^-, Br^-, I^-, and S^{--}. For phosphorus the authors propose the PH_2^- anion as the most probable product formed. These ion formations have been proved to be quantitative by elemental analysis methods, except for phosphorus which is not reported in the literature as having been determinated qualitatively or quantitatively.

The LASSAIGNE test requires the following physical and chemical conditions: a reaction zone temperature of several hundred °C (BUNSEN flame temperature applied to the reagent glass), an alkali metal as the reagent, and an organic compound containing nitrogen, a halogen, or sulfur. All of these conditions are inherent in the *N*-FID where the rubidium bromide surface temperature ranges approximately between 500° and 650°C, the alkali metal atoms are formed by the presence of active hydrogen at the flame temperature, and the nitrogen (halogen, sulfur, phosphorus) is introduced as the glc peak, reacting with the alkali metal atoms to form alkali ions yielding the detector signal.

The reaction rate is a function of the alkali salt used and also a function of the heteroelement. While each single alkali atom reacts with one phosphorus atom, only about every tenth alkali atom reacts with one nitrogen atom, and only about every 1,000th alkali atom reacts with one iodine atom. The halogens fluorine, chlorine, and bromine are suppressed when rubidium bromide is used.

Nitrogen-containing organic compounds of different chemical classes (*i.e.*, primary, secondary, and tertiary amines; amides; azo, nitro, and nitroso compounds; aromatic structures) and amino-acid derivatives have been analyzed by the *N*-FID, and the signal is proportional to the nitrogen content in the molecule. This is in accordance with the quantitative classical microchemical determinations of halogens and sulfur using alkali metals as reactants. The reaction taking place is described by the following equations:

$$RbBr \overset{\Delta}{\rightleftharpoons} Rb^+ + Br^- \overset{H}{\rightleftharpoons} Rb + H^+ + Br^- \quad (2)$$

$$Rb + RCN \rightleftharpoons Rb^+ + CN^- + R \quad (3)$$

$$(R = \text{organic radical})$$

The detector signal (peak) originated from the above reaction is due to the rubidium ion formation in equation (3).

The ionization degree of the *N*-FID is about $10^{-1}\%$ compared to the $10^{-3}\%$ ionization degree of the FID. Thus, the signal enhancement of the *N*-FID is about 100 for nitrogen-containing organic compounds.

The ionization degree of a thermionic phosphorus detector has been shown (Riedmann and Euston 1969) to be linearly related to the value of the background current caused by the ionized rubidium ions in equilibrium with the rubidium atoms:

$$Rb^+ \overset{H}{\rightleftharpoons} Rb \quad (4)$$

This equilibrium is of a constant value over a wide range of hydrogen, air, and carrier flow rates. Thus, the sensitivity is independent of flow settings within certain flow ranges. The signal output, of course, increases with the increase of any flow rate. This is due to the effective increase in temperature (hydrogen, air) and the increase in the area of the rubidium bromide crystal in contact with the flame (carrier gas), both forming a larger amount of rubidium ions and rubidium atoms.

Thermionic detectors can be operated to yield 100% ionization over a short period of time. This period of operation is more of theoretical than of practical value and is not recommended, although selectivity is increased by a factor of about 1,000 over that obtained with the selectivity at recommended operating parameters. The operating parameters of the detector yielding 100% ionization result in a flame temperature causing the crystal to melt and, therefore, to reduce its stability, life time, and low noise characteristics.

When selecting an alkali halide crystal as the source, the detector is insensitive to organochlorine pesticides. This selective suppression is due to the equilibrium of the reaction shown earlier as equation (1) being shifted almost completely to the left in the presence of chlorinated organic compounds. Iodine-, nitrogen-, and phosphorus-containing compounds shift the equilibrium to the right. The value to which the equilibrium is shifted is a function of the difference in ionization (vapour pressure) between the source halide and the alkali salt formed by the reaction between the alkali source atoms and the heteroelement (RIEDMANN 1973). The value of the equilibrium in equation (3), shifted to the formation of rubidium ions, is the sensitivity of the detector for the corresponding heteroelement. This results in differentiation between phosphorus and nitrogen-containing pesticides using rubidium bromide and potassium chloride. Both salts show the same sensitivity for phosphorus, but differ by the factor of about ten in their sensitivity for nitrogen. Potassium chloride shows ten times less sensitivity for nitrogen than rubidium bromide does.

d) Operating parameters

All the parameters inducing a rise in temperature of the crystal surface and causing rubidium bromide to evaporate into the detector flame reaction zone influence the output signal and the selectivity; they include hydrogen flow, air flow, and carrier flow rates, detector base temperature, crystal position, and jet internal diameter.

1. Hydrogen flow rate. — The strongest influence on flame temperature arises from the hydrogen flow rate. Accurate flow settings are not necessary for the design of the thermionic detector described here in order to operate it. This is due to a feature allowing the adjustment of the crystal surface with respect to the flame. The sensitivity of the nitrogen detector is independent of the hydrogen flow rate in the range between 20 and 32 ml/min. At lower flows than about 20 ml/min the nitrogen signal output for practical work is too low with respect to the required selectivity. At flow rates higher than about 32 ml/min the crystal temperature exceeds the melting point of 680°C for rubidium bromide causing a larger increase in the noise relative to the signal output. This noise increase is due to the relatively large changes in the rubidium bromide vapour pressure as a result of flame temperature fluctuations.

In order to keep the nitrogen signal output constant within ±1%, the hydrogen flow rate has to be held stable within ±0.2%. In order to accomplish this flow stability, a BROOKS flow controller is used in series with a needle valve and a high quality head-pressure regulator.

The calculated ratio of active rubidium atoms/rubidium ions in equation (2) is in the range of 10^4. The same ratio was found using potassium chloride in the selective detection of phosphorus *versus* nitrogen. The amount of rubidium atoms actively present in the reaction zone is about 10^{-12} g/sec at the recommended operating parameters.

The detector selectivity is increased by two factors when a high hydrogen flow rate in the range 28 to 30 ml/min is used. First the crystal surface is subjected to a large relative temperature change when approximately one-to-ten μg quantities of hydrocarbon elute into the flame reaction zone, thus causing the background current to decrease by a greater absolute value than at a low (low background current) hydrogen flow. Secondly, at higher hydrogen flows the electrometer range and attenuator have to be set to higher values reducing the hydrocarbon signal with respect to the nitrogen output signal.

2. Air flow rate. — The air flow rate is recommended to be set to the lowest value that will maintain the flame. By reducing the air flow, the detector selectivity is increased by suppressing the hydrocarbon signal. The main reason for the suppression of the hydrocarbon is the lower oxidative character of the FID ionization properties.

In order to keep the nitrogen output signal constant within ±1% the air flow rate has to be held constant within ±2.4 ml/min. A high quality head-pressure regulator in series with a needle valve is sufficient for the required flow stability.

3. Carrier flow rate. — In order to obtain the best separation of a sample, the gas chromatographic column is operated at that carrier gas flow rate which ensures the maximum number of theoretical plates. A carrier gas flow rate between 30 and 50 ml/min is commonly selected. The thermionic nitrogen detector operates best within 60 to 90 ml of helium/min when high output signal and high selectivity are required. To obtain the maximum plate number and the optimum detector performance "make up" helium is added to the detector at the column outlet. The increase of the helium flow rate is limited by the fact that too much helium will possibly blow out the flame when the sample is injected.

The nitrogen output signal is almost linear in the carrier flow range of 50 to 90 ml/min. To ensure a ±1% signal output stability the carrier gas must be kept constant within ±0.4%, e.g., ±0.24 ml/min at 60 ml helium flow.

The nitrogen detector signal output is linearly dependent on the background current resulting from the gas flow parameters discussed above. The background current is an exponential function of the carrier flow. The exponential function can be explained by the size of the area of the crystal in contact with the flame. An increase in the carrier flow causes the flame diameter in the zone of maximum temperature to increase proportionally to the carrier flow rate. Thus, the area of the flame zone in contact with the crystal is proportional to the square of the radius yielding an exponential increase of the amount of salt evaporated from the surface. A cooling effect due to the thermal

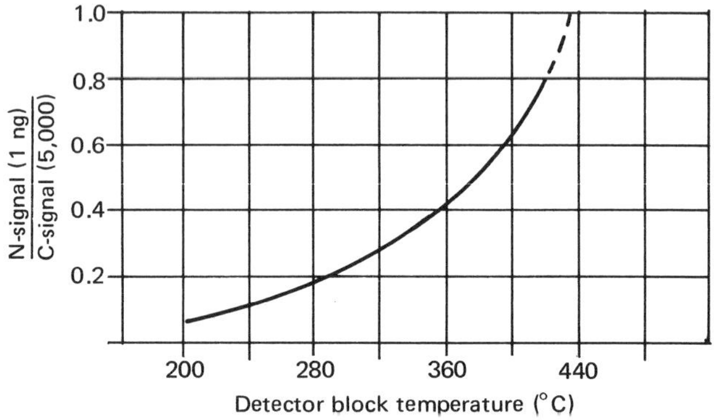

Fig. 5. Effect of detector block temperature on nitrogen/carbon selectivity with crystal in position 3.

conductivity of helium can be neglected with respect to these considerations.

4. Jet diameter. — The internal diameter of the jet influences flame temperature and flame shape. Small internal jet diameters cause a higher flame temperature and a narrower flame shape than larger diameters do at the same gas flow rates. Jets of 0.3 to 0.4 mm id are used for operation within the recommended flow rates.

5. Detector block temperature. — With rise of the detector block temperature the selectivity of the *N*-FID increases due to the larger amount of salt evaporating according to the vapour pressure dependence on the absolute temperature. The nitrogen signal output increases due to this fact, while the hydrocarbon signal is clearly reduced under this condition. Thus, the selectivity increase is due to both factors, the increasing nitrogen signal output and the decreasing carbon signal output. Between 240° and 430°C the selectivity differs by a factor of ten (Fig. 5). Glass column sealing materials such as silicone rings and Teflon ferrules do not survive the required high detector block temperature. A new material, the *Sigri* graphite O-rings, cure the problems of leaking column-detector connections. This material resists temperatures up to 1,200°C in an oxygen-free atmosphere. Columns sealed with graphite O-rings are leak-free over unlimited operation for isothermal or programmed work. Graphite sealing is recommended for electron-capture detector work as well as for all detectors operating at temperatures higher than 200°C when glass columns are used.

6. Detector sensitivity. — The signal-to-noise ratio of the *N*-FID is constant within a wide range of gas flows and detector temperatures. Within these parameters of detector operation the sensitivity cannot be improved, as shown by the equation governing the signal-to-noise ratio of the *N*-FID:

$$Rb^+ \overset{H}{\rightleftharpoons} Rb \qquad (4)$$

The rubidium ions present in the flame correspond to the background current. Background current fluctuation, due to gas flow pulsations, corresponds to the noise. Gas flow pulsations cause a temperature fluctuation of the flame which results in detector noise. Thermionic detector noise is a wandering rather than short-term noise. The rubidium atoms on the right side of equation (4) represent detector signal output and linear dynamic range. As long as the equilibrium is unchanged the signal-to-noise ratio is constant over a wide range of detector operating parameters.

7. Gate electrode. — An increase in the signal-to-noise ratio of thermionic detectors can be improved only by a noise level reduction. The noise of the *N*-FID can be reduced by addition of a gate electrode positioned at the top of the jet tip which is operated with a negative potential variable from zero to —350 V; the jet has a +350 V potential. The operating principle of the gate electrode is based on the compensation of detector noise by distributing the rubidium ions forming the background current into two equivalent parts between collector electrode and gate electrode. The gate electrode is operated at that negative potential which exactly compensates the background current induced by the rubidium ions. A potentiometer is used for the compensation process and the zeroing is recorded. The zeroing affects the noise to be reduced by a factor of ten, thus improving the sensitivity by the same value. With the gate electrode operating the nitrogen peak is inverted, as is the carbon peak (Fig. 6).

III. Use of the *N*-FID in the gas chromatographic residue analysis of nitrogen-containing pesticides

In the pesticide manual (H. Martin 1972) published by the British Crop Protection Council are listed 502 chemicals used as the active components of the pesticides discussed; among these 294 (>58%) are organic nitrogen-contai .ing compounds and an additional 51 (54%) of 94 now superseded nitrogen-containing compounds are currently of little commercial interest. Among the possible temporary residues originating from pesticide use, numerous organic nitrogen-containing compounds, degradation products, and metabolites of all these nitrogen-containing pesticidal active compounds may, therefore, occur in the environment. For pesticide residue analysis 161 of these compounds and three organic phosphorus-containing pesticides were evaluated by us under the following glc conditions:

Instruments:	*Hewlett-Packard*, Models 5700A and 5750G gas chromatographs
Detectors:	FID, *N*-FID

Columns:	Glass, 6 ft. long, 2 mm id.
	a) SE 30, 5% on Chromosorb W-HP, 80/100 mesh
	b) QF 1, 5% on Chromosorb W-HP, 80/100 mesh
Temperatures:	Oven, 220°, 160° and 150°C (isothermal)
	Injection port, 250°C
	Detector, 350°C
Gases:	Carrier gas, helium 40 ml/min
	Fuel gases, hydrogen 30 ml/min, air 220 and 350 ml/min

In Table XX the 164 compounds investigated are numerically arranged according to the alphabetical sequence of their common names and are also characterized by their chemical names. Degradation products included show those active compounds from which they originated. Moreover, the use of the active compounds as fungicide, herbicide, insecticide, acaricide, or nematicide is given. For better comprehension Table XXI shows some trade names of the formulations of the determined active compounds in the alphabetical sequence and indicates which of their degradation products are included in the evaluation.

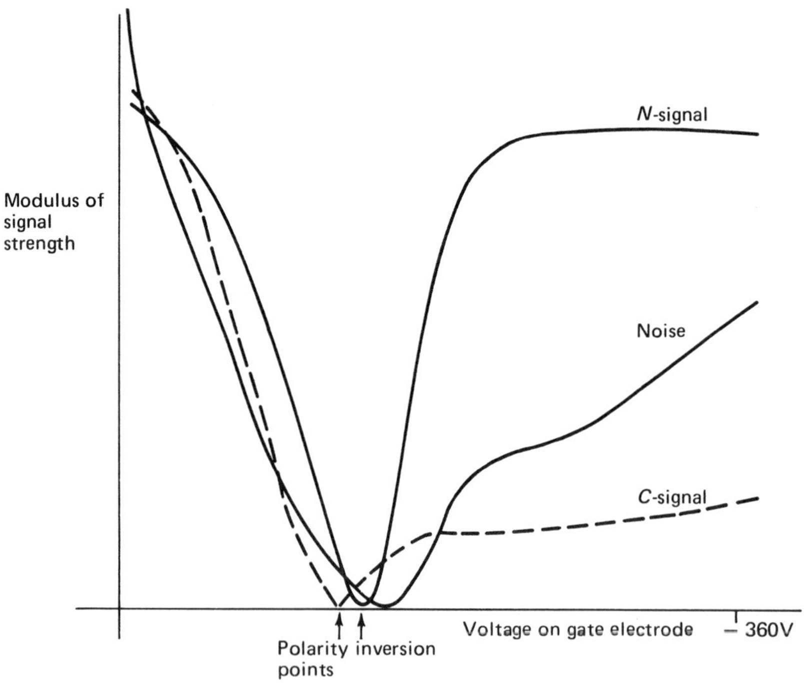

Fig. 6. Effect of gate electrode voltage on sensitivity.

The following paragraphs a) to f) as well as Tables III to V and VII to XIX contain details of the evaluated absolute and relative retention times as well as further information about the glc of active pesticide components among groups of amides, thioamides, nitriles, carbamates, thiocarbamates, urea herbicides, nitrobenzene derivatives, amino compounds, and heterocyclic compounds with one, two, and three ring-nitrogen atoms, and the chemical changes in the molecules resulting from them after application. Most of these compounds can be determined in the pg range (relative to nitrogen) with the *N*-FID.

a) Amides, thioamides, and nitriles

Examples for amides, thioamides, and nitriles which can be determined well with the *N*-FID as per our experience are the herbicides propachlor (147), monalide (119), chlorthiamid (50), dichlobenil (63), and ioxynil (92), as well as the 2,6-dichlorobenzamide (64) originating from chlorthiamid (50) and dichlobenil (63) in soil, plants and animals (BEYNON and WRIGHT 1972, VERLOOP 1972), and the insecticide cartap (33). Table III presents the glc retention times of these compounds. The equality in the retention times of chlorthiamid (50) and dichlobenil (63) under the same glc conditions on the SE 30 column as well as on the QF 1 column results from the change of chlorthiamid (50) into dichlobenil (63) during the analytical process. The relative retention times of 0.08 min on the SE 30 column and of 0.07 min on the QF 1 column, at a column temperature of 160°C, are the retention times of dichlobenil (63). As the change of chlorthiamid into dichlobenil does not quantitatively take place during the analytical process, it is recommended to oxidize chlorthiamid with permanganate to dichlobenil before glc determination (BEYNON *et al.* 1966 a). Dichlobenil (63) and propachlor (147), for example, have also been determined with the COULSON electrolytic conductivity detector (LASKI and WATTS 1973), and chlorthiamid (50), dichlobenil (63), and ioxynil (92) with electron-capture detectors (BEYNON *et al.* 1966 a, and b, MEULEMANS and UPTON 1966). Similar to the nitrophenols, peak tailing occurs with ioxynil in gas chromatography. This can be avoided, as it can with the nitrophenols, by methylation of the hydroxyl group using diazomethane (GUTENMANN and LISK 1965) or by gas chromatography on suitable columns in the presence of orthophosphoric acid (HRIVNÁK and STOTA 1968).

Also, the organophosphorus insecticides derived from the carbonic acid amides and thiamides or from nitriles, acephate (1), methamidophos (2), dimethoate (71), omethoate (72), formothion (90), monocrotophos (120), dicrotophos (67), phosphamidon (142), its metabolite in plants and animals desethylphosphamidon (143) (ANLIKER *et al.* 1961, CLEMONS and MENZER 1968), and phoxim (144) can

Table III. *Retention times of amides, thioamides, and nitriles of structures:*

A: $R—CO—N\langle^{(a)}_{(b)}$ B: $H_2N—(a)$ C: $R—CS—N\langle^{(a)}_{(b)}$ D: $R—CN$

No. in Table XX	Structure	Substituents in the structure: R	Substituents in the structure: (a)	Substituents in the structure: (b)	Compound[a]	Column temp.[b] (°C)	Retention times: Absolute (10^{-2} min): SE 30[c]	Retention times: Absolute (10^{-2} min): QF 1[d]	Retention times: Relative to parathion = 1.00: SE 30[c]	Retention times: Relative to parathion = 1.00: QF 1[d]
147	A	$CH_2Cl—$	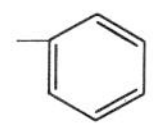	$—CH(CH_3)_2$	propachlor	220 160	84 365	108 621	0.41 0.22	0.35 0.20
119	A	$CH_3—CH_2—CH_2—C(CH_3)_2—$	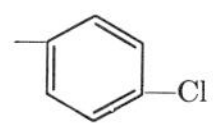	H	monalide	220 160	147 1,092	153 1,149	0.71 0.58	0.49 0.37
33	A	$(CH_3)_2N—CH(CH_2—S—)—CH_2—S—CO—NH_2$	H	H	cartap	150	172	—	0.05	—
1	A	$CH_3—$	$—P(O)(OCH_3)(SCH_3)$	H	acephate	220 160	143 223	— —	0.72 0.14	— —
2	B	—	$—P(O)(OCH_3)(SCH_3)$	—	methamidophos (acephate-met)	160	72	—	0.05	—
71	A	$(CH_3O)_2P(S)—S—CH_2—$	$—CH_3$	H	dimethoate	220 160	105 543	182 1,402	0.51 0.33	0.59 0.45
72	A	$(CH_3O)_2P(O)—S—CH_2—$	$—CH_3$	H	omethoate (dimethoate-met)	220 150	78 700	168 1,332	0.38 0.20	0.55 0.28
90	A	$(CH_3O)_2P(S)—S—CH_2—$	$—CH_3$	—CHO	formothion	220 160	139 874	251 2,365	0.67 0.53	0.81 0.77
120	A	$(CH_3O)_2P(O)—O—C(CH_3)=CH—$	$—CH_3$	H	monocrotophos	220 160	90 466	210 1,973	0.43 0.28	0.68 0.64
67	A	$(CH_3O)_2P(O)—O—C(CH_3)=CH—$	$—CH_3$	$—CH_3$	dicrotophos	220 160	88 —	208 1,909	0.42 —	0.66 0.62
142	A	$(CH_3O)_2P(O)—O—C(CH_3)=CCl—$	$—C_2H_5$	$—C_2H_5$	phosphamidon	220 160 150	124, 143 803, 1,042 1,670, 2,198	269, 376 2,878 —	0.60, 0.71 0.49, 0.64 0.48, 0.63	0.88, 1.24 0.89 —

143	A	$(CH_3O)_2P(O)—O—C(CH_3)=CCl—$	$—C_2H_5$	H	phosphamidon-met	220 160 150	97, 122 666, 795 1,785	244 2,485 —	0.47, 0.59 0.41, 0.49 0.51	0.79 0.81 —
50	C	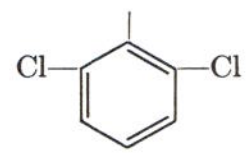	H	H	chlorthiamid	160 150	135 275	206 295	0.08 0.08	0.07 0.06
63	D	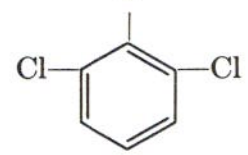	—	—	dichlobenil	220 160	52 134	63 206	0.25 0.08	0.20 0.07
64	A	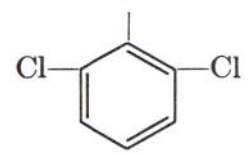	H	H	dichlobenil-met	220 160	91 437	143 915	0.43 0.27	0.46 0.30
92	D	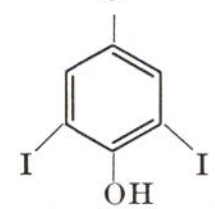	—	—	ioxynil	220 160	— —	213 1,454	— —	0.69 0.47
144	D	$(C_2H_5O)_2P(S)—O—N=C—$ 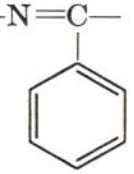	—	—	phoxim	150	160, 237	—	0.04, 0.06	—

[a] Pesticides with "-met" added to their names are not invariable pesticides, but metabolites, degradation products, or contaminations.
[b] 6 × 2 mm id. glass column.
[c] Column packed with 5% SE 30 on 80/100 mesh Chromosorb W (HP).
[d] Column packed with 5% QF 1 on 80/100 mesh Chromosorb W (HP).

be determined with the *N*-FID in glc. Table III contains their glc retention times under the given operating conditions.

Among 95 nitrogen-containing organic pesticides which can be determined with the COULSON electrolytic conductivity detector, LASKI and WATTS (1973) denominated dimethoate (71), omethoate (72), monocrotophos (Azodrin®, 120), dicrotophos (Bidrin®, 67) and phosphamidon (142). As is known, the insecticide omethoate (72) is not only applied as pesticide, but also occurs as a metabolite of dimethoate (71) in soil (BACHE and LISK 1966) as well as in plants and animals (O'BRIEN 1960, HACSKAYLO and BULL 1963, LUCIER and MENZER 1968 and 1970). The relative retention times of monocrotophos and dicrotophos (0.38 and 0.36 min, respectively) are as close to each other in the glc data of LASKI and WATTS (1973) as they are in our Table III. Monocrotophos is not only applied as insecticide but also occurs as a metabolite of dicrotophos in plants and animals (MENZER and CASIDA 1965).

The organophosphorus insecticides in Table III have also been determined and separated by glc with potassium chloride thermionic detectors (IVES and GIUFFRIDA 1967, THORNTON and ANDERSON 1968, STORHERR *et al.* 1971, *Deutsche Forschungsgemeinschaft* 1972, and others), with potassium borate thermionic detectors (GIANG and BECKMAN 1968, BECKMAN and GARBER 1969), with caesium bromide thermionic detectors (RUZICKA *et al.* 1967, EBING 1968), as well as with flame photometric detectors (BOWMAN *et al.* 1967 and 1971, STORHERR *et al.* 1971, VOSS *et al.* 1971), and with electron-capture detectors (MIDDELEM and WAITES 1964, BURKF and HOLSWADE 1966, and others).

b) *Carbamates and thiocarbamates*

The *N*-FID has been proved in the glc analysis of carbamates and thiocarbamates. Table IV lists the retention times determined under the given operating conditions of the carbamate insecticides carbaryl (32), propoxur (150), dioxacarb (76), aminocarb (7), butoxicarboxim (29), mercaptodimethur (98) and of its metabolites in plants (BOWMAN and BEROZA 1969), mercaptodimethursulfoxide (99) and mercaptodimethursulfone (100), of the carbamate herbicides propham (149), chlorpropham (49), chlorbufam (39), barban (22), phenmedipham (137), and of the thiocarbamate herbicides di-allate (59), triallate (160), butylate (30), and cycloate (55).

Propham (149) and chlorpropham (49), but not all carbamates, can be gas chromatographed without decomposition (KRISHNA *et al.* 1962, ZIELINSKI and FISHBEIN 1965, PECK and HARKISS 1971). Those carbamates that partially, but not completely, decompose during the glc process, which may be indicated by several peaks with different retention times during the analysis, cannot be quantitatively determined with the necessary certainty in this manner. Examples for this are mercaptodimethur (98) and dioxacarb (76) in Table IV. It is therefore suitable to convert them into glc stable derivatives, e.g., mercaptodimethur (98) and its metabolites (99 and 100) by hydrolysis into the corresponding phenols, which then can be gas chromatographed by means of, for instance, a flame photometric detector sensitive to sulfur (BOWMAN and BEROZA 1969).

The azo compounds 138, 139, and 140 listed in Table IV were included in the evaluation with regard to the questions whether such compounds are present in food, are analytically determinable, and are possibly of significance to the toxicological valuation. Azo compounds have been reported as oxidation products of derivatives of aromatic amines used as pesticides under experimental conditions in soil (see MAIER-BODE 1971, pp. 178–180; GEISSBÜHLER and VOSS 1971). The compounds 138 (3,3′-dimethylazobenzene), 139 (3,3′-dihydroxyazobenzene), and 140 (3-methyl-3′-hydroxyazobenzene) might occur as metabolites of the herbicide phenmedipham (137). They can be gas chromatographed very well with the *N*-FID. In food we never did find them as residues of phenmedipham. Even other aromatic azo compounds can be as well detected with the *N*-FID with high sensitivity, e.g., compounds 38, 54, 86, 97, 111, and 125 (Table V) and the azometabolite 3,3′-dichloroazobenzene originating from the herbicide carbamates chlorpropham (49), chlorbufam (39), or barban (22).

The COULSON electrolytic conductivity detector also served for the detection of carbaryl (32), propoxur (150), aminocarb (7), mercaptodimethur (98), propham (149), chlorpropham (49), barban (22), phenmedipham (137), di-allate (59), tri-allate (160), butylate (30), and cycloate (55) (LASKI and WATTS 1973). Furthermore, flame ionization detectors were used in the glc analysis of carbaryl (32), propoxur (150), propham (149), and chlorpropham (49) (WHEELER and STROTHER 1966, PECK and HARKISS 1971) and electron-capture detectors were used for chlorpropham (49) (MENGES and HUBHARD 1970) and tri-allate (160) (MCKONE and HANCE 1967) as well as for carbaryl (32), propoxur (150), and aminocarb (7) after alkaline hydrolysis and reaction of the formed amines with 1-fluro-2,4-dinitrobenzene to the corresponding dinitroaniline derivatives (HOLDEN *et al.* 1969) and for phenmedipham (137) after the alkaline hydrolysis to 3-methylaniline and bromination to 2,4,6-tribromo-3-methylaniline (KOSSMANN 1970).

c) *Urea herbicides*

A number of urea herbicides and many of their degradation products (GEISSBÜHLER and VOSS 1971) are determinable by glc. Table V contains the glc retention times of the 14 herbicides fluometuron

Table IV. *Retention times of carbamates and thiocarbamates of structures:*

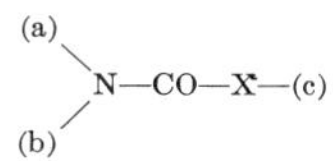

A

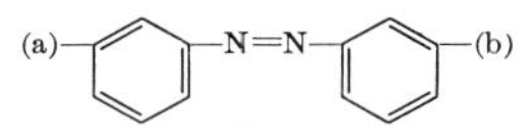

B

No. in Table XX	Structure	Substituents in the structure				Compound[a]	Column temp.[b] (°C)	Retention times			
								Absolute (10^{-2} min)		Relative to parathion = 1.00	
		(a)	(b)	X	(c)			SE 30[c]	QF 1[d]	SE 30[c]	QF 1[d]
150	A	CH_3—	H	O	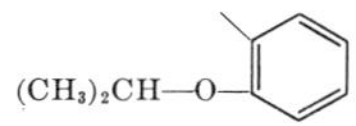	propoxur	160	354	584	0.22	0.19
98	A	CH_3—	H	O	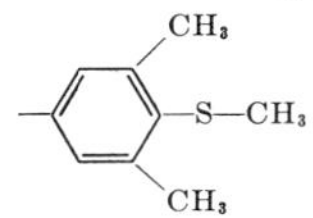	mercaptodimethur	220 160	73 306, 1,455	187 197, 1,635	0.35 0.18, 0.85	0.60 0.06, 0.53
99	A	CH_3—	H	O	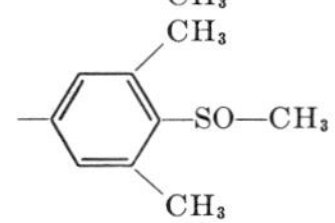	mercaptodimethur-met	220 160	160 1,283	296 2,752	0.77 0.75	0.96 0.89
100	A	CH_3—	H	O	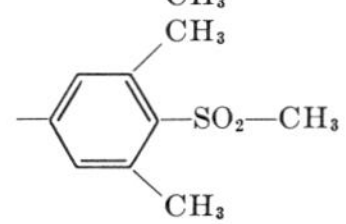	mercaptodimethur-met	220 160	170 1,379	324 3,198	0.82 0.81	1.07 0.99
7	A	CH_3—	H	O	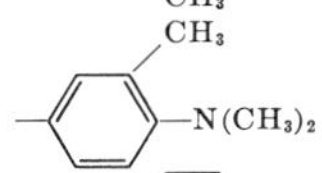	aminocarb	220 160	114 742	124 893	0.55 0.44	0.40 0.29
32	A	CH_3—	H	O	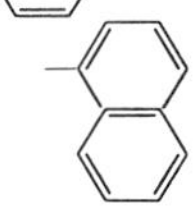	carbaryl	220 160	69 193	63 210	0.32 0.12	0.20 0.07

76	A	CH_3-	H	O	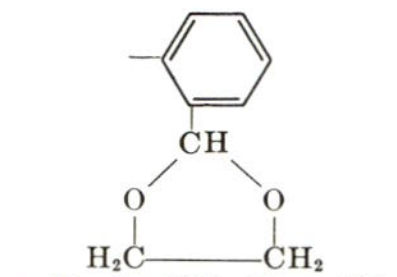	dioxacarb	160	175, 965	201, 2,152	0.10, 0.59	0.07, 0.70
29	A	CH_3-	H		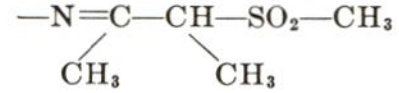	butoxicarboxim	160	160	—	0.10	—
149	A	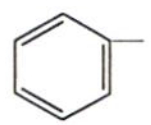	H	O	$-CH(CH_3)_2$	propham	160 150	192 427	199 345	0.12 0.11	0.06 0.07
49	A	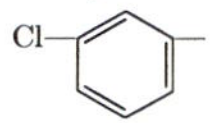	H	O	$-CH(CH_3)_2$	chlorpropham	220 160	88 432	83 419	0.42 0.26	0.27 0.14
39	A	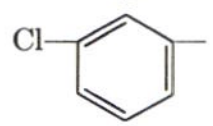	H	O	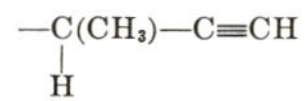	chlorbufam	200 100	105 258	90 523	0.50 0.16	0.29 0.17
22	A	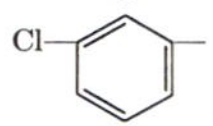	H	O	$-CH_2-C\equiv C-CH_2Cl$	barban	220 160	328 3,578	273 2,974	1.54 2.17	0.89 0.97
137	A	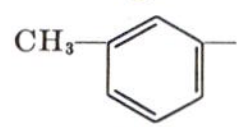	H	O	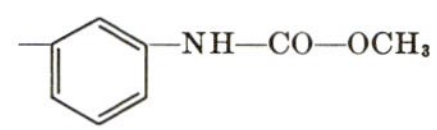	phenmedipham	220 160	181 —	143 906	0.87 —	0.46 0.29
138	B	CH_3-	$-CH_3$	—	—	phenmedipham-met	220 160	151 1,111	97 554	0.73 0.66	0.31 0.18
139	B	HO—	—OH	—	—	phenmedipham-met	220 160	461 —	391 4,904	2.23 —	1.29 1.52
140	B	CH_3-	—OH	—	—	phenmedipham-met	220 160	256 2,380	185 1,563	1.24 1.44	0.60 0.51
59	A	$(CH_3)_2CH-$	$(CH_3)_2CH-$	S	$-CH_2-CCl{=}CHCl$	di-allate	220 160 150	— 554 1,153, 1,217	74 301, 351 532, 617	— 0.34 0.31, 0.32	0.24 0.10, 0.12 0.11, 0.13
160	A	$(CH_3)_2CH-$	$(CH_3)_2CH-$	S	$-CH_2-CCl{=}CCl_2$	tri-allate	220 160	142 904	88 469	0.69 0.55	0.28 0.15
30	A	$(CH_3)_2CH-CH_2-$	$(CH_3)_2CH-CH_2-$	S	$-C_2H_5$	butylate	160 150	265 402	123 208	0.16 0.11	0.04 0.04
55	A	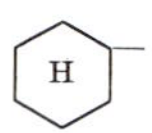	C_2H_5-	S	$-C_2H_5$	cycloate	220 160 150	90 413 898	64 257 483	0.43 0.25 0.24	0.21 0.08 0.10

[a–d] See footnotes to Table III.

Table V. *Retention times of urea derivatives of structures:*

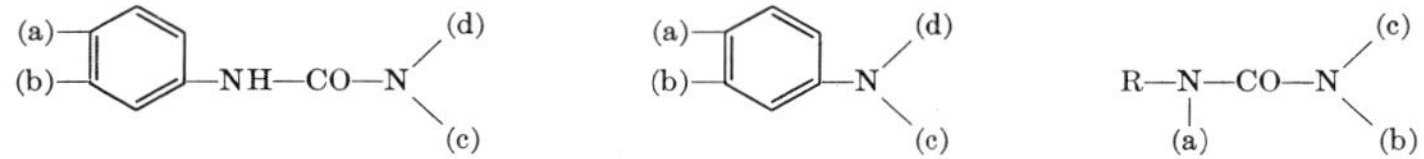

A B C

No. in Table XX	Structure	Substituents in the structure					Compound[a]	Column temp.[b] (°C)	Retention times			
									Absolute (10^{-2} min)		Relative to parathion = 1.00	
		R	(a)	(b)	(c)	(d)			SE 30[c]	QF 1[d]	SE 30[c]	QF 1[d]
82	A	—	H	CF_3—	—CH_3	—CH_3	fluometuron	160 150	425 95	— 58	0.25 0.02	— 0.01
83	A	—	H	CF_3—	H	—CH_3	fluometuron-met	220 160	— 558	149 1,202	— 0.34	0.48 0.39
84	A	—	H	CF_3—	H	H	fluometuron-met	220 160	— 595	162 1,323	— 0.35	0.52 0.49
85	B	—	H	CF_3—	H	H	fluometuron-met	160 150	— 108	58 72	— 0.03	0.02 0.01
86	B	—	H	CF_3—	=N—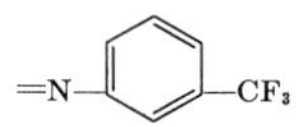—CF_3		fluometuron-met	220 160 150	74 333 702	65 277 428	0.36 0.20 0.19	0.21 0.09 0.09
77	A	—	Cl	Cl	—CH_3	—CH_3	diuron	160 150	127 260	108 152	0.08 0.07	0.03 0.03
126	A	—	Cl	Cl	—CH_3	—C_4H_9(n)	neburon	160 150	141 258	106 152	0.07 0.07	0.03 0.03
51	A	—	CH_3—	Cl	—CH_3	—CH_3	chlortoluron	150	207	115	0.05	0.02
52	A	—	CH_3—	Cl	H	—CH_3	chlortoluron-met	150	215	117	0.06	0.02
53	B	—	CH_3—	Cl	H	H	chlortoluron-met	160 150	114, 166 238, 347	104, 134 150, 228	0.07, 0.10 0.06, 0.09	0.03, 0.04 0.03, 0.05
54	B	—	CH_3—	Cl	=N—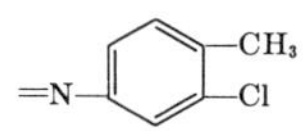 (CH_3, Cl)		chlortoluron-met	220 160	473 —	254 2,406	2.28 —	0.82 0.78
112	A	—	CH_3O—	Cl	—CH_3	—CH_3	metoxuron	160 150	183 355	170 240	0.11 0.09	0.06 0.05
113	A	—	CH_3O—	Cl	H	—CH_3	metoxuron-met	160 150	167 367	168 247	0.10 0.10	0.05 0.05
114	B	—	CH_3O—	Cl	H	H	metoxuron-met	160 150	188, 295 390, 611	206, 296 323, 550	0.11, 0.17 0.10, 0.16	0.07, 0.10 0.07, 0.12
46	A	—	Cl—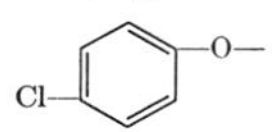—O—	H	—CH_3	—CH_3	chloroxuron	220 160	179 1,262	135 913	0.86 0.77	0.44 0.30

47	A	—	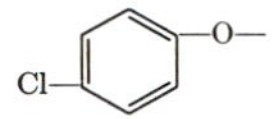	H	H	H	chloroxuron-met	220	193	149	0.93	0.48
48	B	—	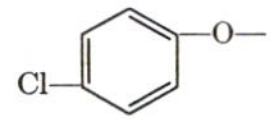	H	H	H	chloroxuron-met	220 160	190, 256 1,378, 2,231	145, 182 1,052, 1,571	0.92, 1.24 0.84, 1.36	0.47, 0.59 0.34, 0.51
28	A	—	Cl	H	$—CH_3$	—CH—C≡CH	buturon	220 160	306 —	545 6,501	1.48 —	1.74 2.02
121	A	—	Cl	H	$—OCH_3$	$—CH_3$	monolinuron	220 160	113 699	143 1,009	0.55 0.41	0.46 0.33
122	A	—	Cl	H	H	$—CH_3$	monolinuron-met	160	1,271	2,290	0.75	0.74
123	A	—	Cl	H	$—OCH_3$	H	monolinuron-met	160	845	1,475	0.50	0.48
124	B	—	Cl	H	H	H	monolinuron-met	160	95, 133	87, 107	0.05, 0.08	0.03, 0.04
125	B	—	Cl	H	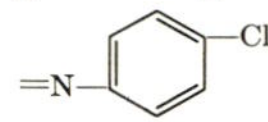		monolinuron-met	220 160	227 2,006	— —	1.10 1.07	— —
107	A	—	Br	H	$—OCH_3$	$—CH_3$	metobromuron	220 160	146 995	181 1,440	0.71 0.59	0.58 0.47
108	A	—	Br	H	H	$—CH_3$	metobromuron-met	160	1,968	3,379	1.16	1.05
109	A	—	Br	H	H	H	metobromuron-met	160 150	135 267	120 178	0.08 0.07	0.04 0.04
110	B	—	Br	H	H	H	metobromuron-met	160 150	127, 180 242, 342	115, 141 167, 227	0.07, 0.10 0.06, 0.09	0.04, 0.05 0.03, 0.05
111	B	—	Br	H	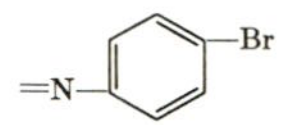		metobromuron-met	220 160	412 —	215 1,773	1.99 —	0.70 0.57
94	A	—	Cl	Cl	$—OCH_3$	$—CH_3$	linuron	220 160	192 1,453	247 2,276	0.93 0.88	0.80 0.74
95	A	—	Cl	Cl	H	$—CH_3$	linuron-met	160 150	131 268	106 283	0.08 0.07	0.03 0.06
96	B	—	Cl	Cl	H	H	linuron-met	160 150	185, 239 390, 493	183 270, 318	0.11, 0.14 0.10, 0.13	0.06 0.06, 0.07
97	B	—	Cl	Cl	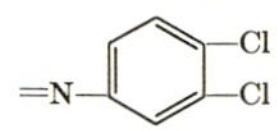		linuron-met	220 160	695, 1,051 —	355, 635 3,703	3.36, 5.10 —	1.15, 2.06 1.20
35	A	—	Br	Cl	$—OCH_3$	$—CH_3$	chlorbromuron	220 160	252 2,127	— —	1.22 1.29	— —
36	A	—	Br	Cl	H	H	chlorbromuron-met	220 160 150	73 — —	71 269 402	0.35 — —	0.23 0.09 0.09
37	B	—	Br	Cl	H	H	chlorbromuron-met	220 160 150	71 — 567	68 253 380	0.34 — 0.15	0.22 0.08 0.08

Table V. (*Continued*)

No. in Table XX	Structure	Substituents in the structure: R	(a)	(b)	(c)	(d)	Compound[a]	Column temp.[b] (°C)	Retention times: Absolute (10^{-2} min): SE 30[c]	QF 1[d]	Relative to parathion = 1.00: SE 30[c]	QF 1[d]
38	B	—	Br	Cl	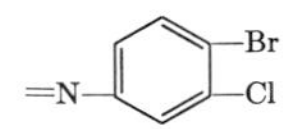		chlorbromuron-met	220	1,293	611	6.25	1.95
56	C	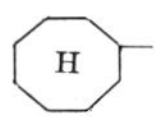	H	CH_3—	—CH_3	—	cycluron	220 160	121 698	142 1,066	0.58 0.42	0.46 0.35
23	C	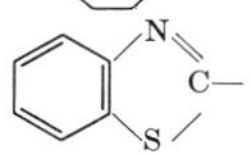	H	H	—CH_3	—	benzthiazuron	220	1,083	653	5.23	2.08
101	C	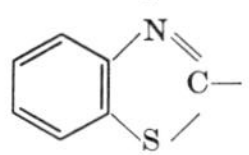	CH_3—	H	—CH_3	—	methabenzthiazuron	220 160	94 463	86 413	0.45 0.27	0.28 0.13

[a–d] See footnotes to Table III.

(82), diuron (77), neburon (126), chlortoluron (51), metoxuron (112), chloroxuron (46), buturon (28), monolinuron (121), metabromuron (107), linuron (94), chlorbromuron (35), cycluron (56), benthiazuron (23) and methabenzthiazuron (101). The *N*-FID is appropriate for their detection. Other detectors were used as well, for instance flame ionization detectors (KATZ and FASSBENDER 1968, KATZ and STRUSZ 1968 and 1969), COULSON electrolytic conductivity detector (LASKI and WATTS 1973), microcoulometric detectors (BURKE and HOLSWADE 1966), and electron-capture detectors (MCKONE and HANCE 1968, MCKONE 1969). However, "gas chromatography of the intact urea herbicides and their major metabolites is still a critical procedure, because of the frequently observed decomposition of these substances in certain chromatographic columns" (VOSS and GEISSBÜHLER 1971). Thermal decompositions have been observed in glc especially with glass columns, less with stainless steel columns (MCKONE and HANCE 1968, SPENGLER and HAMROLL 1970).

For the toxicological evaluation of urea herbicide residues in food it is necessary to consider that some of their degradation products are more or essentially more toxic than the parent herbicides themselves. This is, for instance, valid for the metabolites of fluometuron (82), chlortoluron (51), metabromuron (107), and chlorbromuron (35), which are formed as per Figure 7 and which are shown in

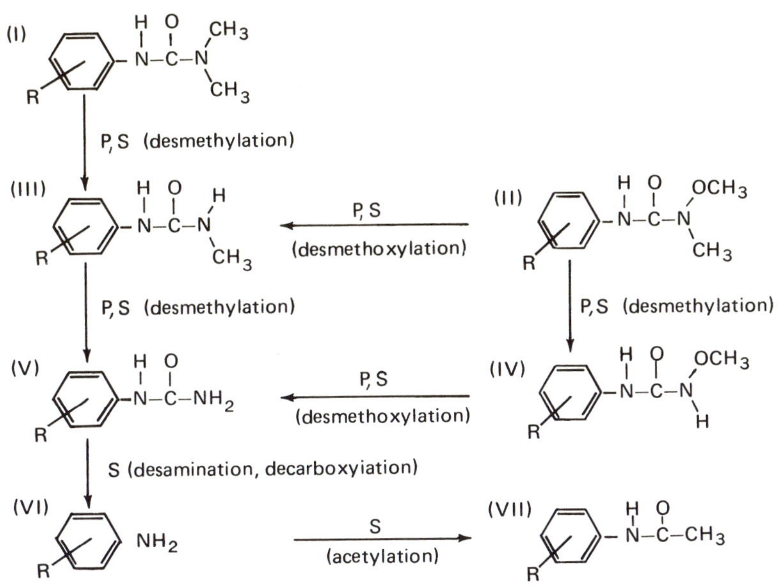

Fig. 7. Metabolic pathways for dimethylphenylurea and methylmethoxyphenylurea herbicides in plants (P) and soils (S) (VOSS and GEISSBÜHLER 1971).

Table VI. *Toxicities as acute oral* LD_{50}*-values to rats of several urea herbicides and some of their metabolites* (VOSS and GEISSBÜHLER 1971).

No. in Table XX	Compound	No. in Fig. 7	LD_{50} (mg/kg)
82	fluometuron	I	>8,000
83	fluometuron-met	III	1,750
84	fluometuron-met	V	1,800
85	fluometuron-met	VI	330
51	chlortoluron	I	>10,000
53	chlortoluron-met	VI	420
107	metabromuron	II	3,000
110	metabromuron-met	VI	410
35	chlorbromuron	II	>5,000
37	chlorbromuron-met	VI	610

Table VI. The aromatic amines 85, 53, 110, and 37 have the highest toxicities.

In case the urea herbicide residues are to be toxicologically evaluated their analyses must not be limited to the determination of unchanged herbicides. In order to be able to include their toxic degradation products in the results of the residue analyses and, simultaneously, to determine the herbicide quantities adsorbed by biological material or by the soil (GEISSBÜHLER and VOSS 1971), it is recommended to hydrolyze the whole herbicide residue with alkali in the presence of the biological material or soil into the corresponding amine, e.g., diuron (77) into 3,4-dichloraniline (96). Then the amines can be analytically determined in different ways, e.g., colorimetrically after diazotation and conversion into azo dyes or, using glc, by means of an electron-capture detector after exchanging the amino groups by the SANDMEYER iodination reaction against iodine atoms (BAUNOK and GEISSBÜHLER 1968, GEISSBÜHLER and VOSS 1971). Besides the urea herbicides 82, 77, 126, 51, 112, 46, 28, 121, 107, 94, 35, and 56 mentioned in Table V, in this way their degradation products 83, 84, 85, 96, 52, 53, 113, 114, 47, 48, 122, 123, 124, 108, 109, 110, 95, 36, and 37 being formed as per Figure 7 are included in the results of the residue analyses.

The azo compounds 86, 54, 125, 111, 97, and 38 mentioned in Table V can be considered as oxidation products of the aromatic amines 85, 53, 124, 110, 96, and 37 and, thus, as the potential metabolites of the urea herbicides fluometuron (82), chlortoluron (51), monolinuron (121), metabromuron (107), linuron (94), and chlorbromuron (35) in soils (EBING 1973). They can be very well evaluated with the *N*-FID in glc. Already GEISSBÜHLER and VOSS (1971) pointed out that extensive analyses by BELASCO and PEASE, by MAIER-BODE, and by themselves did not reveal detectable residues of azobenzenes in a large number of soil samples that had been treated with different urea herbicides under field conditions. In laboratory experiments GEISSBÜHLER and VOSS administered high concentrations of ^{14}C-labelled fluometuron and metobromuron to strongly irradiated cotton and tobacco plants but were not able to detect labelled azobenzene derivatives. From the data at present available they therefore concluded that azobenzenes do not represent significant terminal residues of urea herbicides in either soils or plants.

d) Miscellaneous derivatives of amines

Table VII contains details of glc retention times of the insecticides and acaricides dimefox (70), chlordimeform (40), and chlorfensulphide (45), of the nematicide phenamiphos (134), of the fungicides dichlofluanid (65) and thiophanate (159) and their degradation products 135 and 136 (WAGGONER 1972), 66 (VOGELER and GOELDNER 1967), 41, 42, and 43 (KOSSMANN *et al.* 1971), the theoretical (SCHERING AG 1970) chlordimeform-metabolite 44, which is as per our experience analytically not detectable under practical conditions as a residue in crops, as well as of the quintozene-metabolite pentachloroaniline (153) (GORBACH and WAGNER 1967). The *N*-FID has proved to be appropriate for the glc residue analysis of these compounds. Other detectors used are, for instance, the phosphorus detector described by HARTMANN (1966) for dimefox (70) (RUZICKA *et al.* 1967), the flame photometric detector (BOWMAN *et al.* 1971) or the electron-capture detector (LEMPERLE and STRECKER 1971) for dichlofluanid (65), and the flame ionization detector (KOSSMANN *et al.* 1971), the electron-capture detector (GEISSBÜHLER *et al.* 1971), or the COULSON electrolytic conductivity detector in the nitrogen mode (LASKI and WATTS 1973) for chlordimeform (40).

e) Nitrobenzene derivatives

As an example for a chlorinated nitrobenzene derivative we chose the fungicide quintozene (152). Small quantities of this compound (in the pg range) and of its metabolite pentachloroaniline (153) in biological material and soils can be very well determined by glc because of their high chlorine contents with detectors being sensitive to halogens, e.g., with electron-capture detectors or microcoulometric detectors (BONELLI *et al.* 1964, BURKE and HOLSWADE 1966, GORBACH and WAGNER 1967, METHRATTA *et al.* 1967, KUCHAR *et al.* 1969, and others), and as per our experience also with detectors sensitive to nitrogen, e.g., the *N*-FID. Table VIII shows the glc retention times of quintozene (152), with Table VII including those of pentachloroaniline (153) under the given operating conditions. An example for a nitraniline derivative determinable by glc with the *N*-FID is the herbicide nitralin (127) listed in Table VIII. Derivative formation, e.g., with bis-silyltrifluoracetamide (BSTFA) is recommended.

Table VII. *Retention times of derivatives of miscellaneous amino compounds of structures:*

A: R—N(a)(b) B: R—N=R_1

No. in Table XX	Structure	Substituents in the structure				Compound[a]	Column temp.[b] (°C)	Retention times			
								Absolute (10^{-2} min)		Relative to parathion = 1.00	
		R	(a)	(b)	R_1			SE 30[c]	QF 1[d]	SE 30[c]	QF 1[d]
70	A	$(CH_3)_2N—P(O)(F)—$	$—CH_3$	$—CH_3$	—	dimefox	220 160	49 94	— —	0.25 0.06	— —
134	A	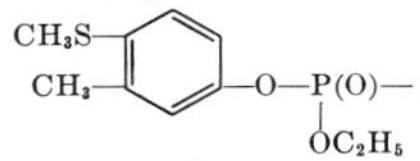	$—CH(CH_3)_2$	H	—	phenamiphos	220 160	314 —	414 5,702	1.52 —	1.36 1.77
135	A	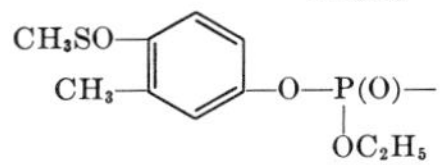	$—CH(CH_3)_2$	H	—	phenamiphos-met	220	685	2,799	3.31	9.03
136	A	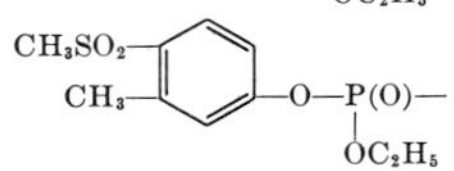	$—CH(CH_3)_2$	H	—	phenamiphos-met	220	694	3,152	3.35	10.17
65	A	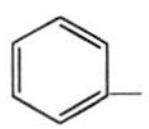	$—SO_2—N(CH_3)_2$	$—S—C(F)Cl_2$	—	dichlofluanid	220 160	194 -,421	255 2,370	0.91 0.86	0.83 0.77
66	A	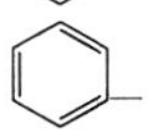	$—SO_2—N(CH_3)_2$	H	—	dichlofluanid-met	220 160	89 414	158 1,093	0.42 0.25	0.51 0.36
159	A	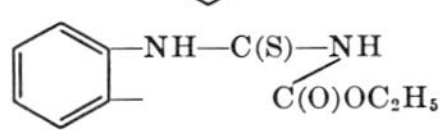	$—C(S)—NH—C(O)OC_2H_5$	H	—	thiophanate	150	105	63	0.03	0.01
40	B	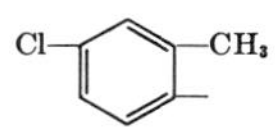	—	—	$=CH—N(CH_3)_2$	chlordimeform	220 160	94 622	82 387	0.44 0.38	0.27 0.13

41	B	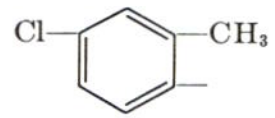	—	—	=CH—NH(CH_3)	chlordimeform-met	220 160	— 448	84 402	— 0.27	0.27 0.13
42	A	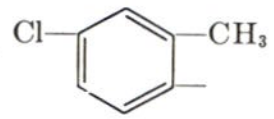	—CHO	H	—	chlordimeform-met	220 160	— 317	165 669	— 0.19	0.37 0.22
43	A	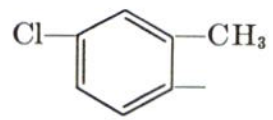	H	H	—	chlordimeform-met	160 150	116 142, 312	116 162, 197	0.07 0.06, 0.08	0.04 0.03, 0.04
44	B	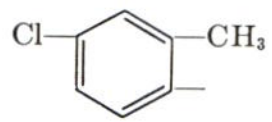	—	—	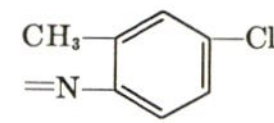	chlordimeform-met	220 160	387 —	207 1,815	1.82 —	0.67 0.59
45	B	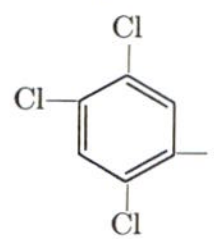	—	—	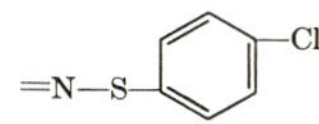	chlorfensul-phide	220 160	376, 485 —	208, 255 1,739, 2,332	1.77, 2.28 —	0.67, 0.82 0.57, 0.76
153	A	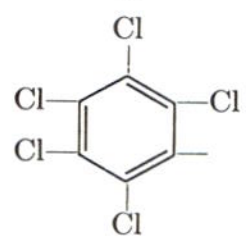	H	H	—	quintozene-met	220 160	155 923	110 576	0.76 0.56	0.36 0.19

[a–d] See footnotes to Table III.

Table VIII. *Retention times of nitrobenzene derivatives of structure:*

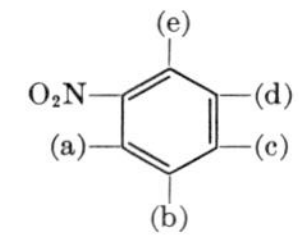

No in Table XX	Substituents in the structure					Compound[a]	Column temp.[b] (°C)	Retention times			
								Absolute (10^{-2} min)		Relative to parathion = 1.00	
	(a)	(b)	(c)	(d)	(e)			SE 30[c]	QF 1[d]	SE 30[c]	QF 1[d]
152	Cl	Cl	Cl	Cl	Cl	quintozene	220 160	127 682	107 549	0.62 0.42	0.35 0.18
78	H	$—NO_2$	—OH	$—CH_3$	H	DNOC	220 160	89 367	137 809	0.43 0.22	0.44 0.26
74	H	$—NO_2$	—OH	$—CH—C_2H_5$ $\vert$ CH_3	H	dinoseb	220 160	139 825	207 1,575	0.67 0.50	0.66 0.51
75	H	$—NO_2$	$—O—CO—CH_3$	$—CH—C_2H_5$ $\vert$ CH_3	H	dinoseb acetate	220 160	175 1,317	374 4,024	0.85 0.80	1.21 1.25
24	H	$—NO_2$	$—O—CO—CH{=}C(CH_3)_2$	$—CH—C_2H_5$ $\vert$ CH_3	H	binapacryl	220	406	719	1.91	2.33
130	H	H	$—O—P(S)(OC_2H_5)_2$	H	H	parathion	220 160 150	206, 209, 211 1,642, 1,700, 1,888 3,471, 3,722	304, 309, 314 3,084, 3,125, 3,225 4,647, 4,720	1.00 1.00 1.00	1.00 1.00 1.00
131	H	H	$—O—P(O)(OC_2H_5)_2$	H	H	parathion-met (paraoxon)	220 160	165 1,385	388 4,572	0.80 0.81	1.28 1.42
132	H	H	$—O—P(S)(OCH_3)_2$	H	H	parathion methyl	220 160	159 1,207	253 2,203	0.77 0.64	0.82 0.71
133	H	H	$—O—P(O)(OCH_3)_2$	H	H	parathion methyl-met (paraoxon-methyl)	220 160	126 879	319 3,139	0.61 0.47	1.03 0.97
81	$CH_3—$	H	$—O—P(S)(OCH_3)_2$	H	H	fenitrothion	220 160	183 1,352	271 2,595	0.89 0.82	0.87 0.85
26	H	$—NO_2$	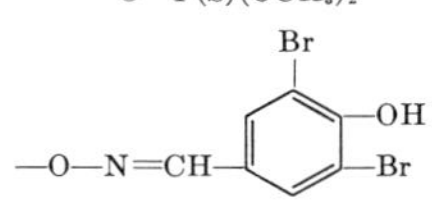 	H	H	bromofenoxim	160	1,657	639	1.01	0.21

27	H	$-NO_2$	$-OH$	H	H	bromofenoxim-met (2,4-dinitrophenol)	220 160	— —	108 548	— —	0.35 0.18
87	H	H	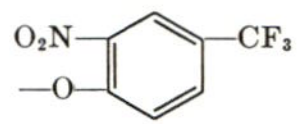	H	H	fluorodifen	220	313	995	1.51	3.27
88	H	H	$-OH$	H	H	fluorodifen-met (4-nitrophenol)	220 160	67 263	87 373	0.32 0.15	0.28 0.17
127	$(CH_3-CH_2-CH_2)_2N-$	$-NO_2$	H	$-SO_2-CH_3$	H	nitralin	220	657	2,388	3.17	7.70

[a–d] See footnotes to Table III.

Free nitrophenols occur as degradation products of pesticides. They are often not very persistent, for they are enzymatically reduced to the corresponding aminophenols under biological conditions, as with 4-nitrophenol in warm-blooded organisms (WILLIAMS 1959) or in sea-water (GRAETZ *et al.* 1970). Parathion (130), parathion methyl (132), and fluorodifen (87) can be degraded to 4-nitrophenol (88), fenitrothion (81) to 3-methyl-4-nitrophenol, and bromofenoxim (26) to 2,4-dinitrophenol (27). These nitrophenols having been formed by the degradation of pesticides as well as other free nitrophenols being applied in this form or as salts in plant protection, such as DNOC (78) and dinoseb (74), may temporarily occur as residues in biological material in soil and in water.

Peak tailing occurs in the glc of nitrophenols. The tailing can be eliminated by methylation of the nitrophenols with diazomethane (BOGGS 1966, YIP and HOWARD 1968, MCKELLAR 1971). The nitrophenol-methyl ethers can be detected with the *N*-FID or with electron-capture detectors and can be separated by glc fron the methylation products of other possible pesticides, for instance, from chlorphenoxy acids (HOWARD and YIP 1971). Peak tailing of the free phenols can also be avoided by using supports coated with polar phases and orthophosphoric acid (HRIVNÁK and STOTA 1968).

Table VIII shows the glc retention times of μg quantities of the free nitrophenols 4-nitrophenol (88), 2,4-dinitrophenol (27), DNOC (78), dinoseb (74), and bromofenoxim (26) (also see LASKI and WATTS 1973). Table VIII also shows the retention times of the nitrophenol ethers and esters fluorodifen (97), dinoseb acetate (75), binapacryl (24), parathion (130), parathion methyl (132), paraoxon (131), paraoxon methyl (133), and fenitrothion (81). Table VIII indicates several values, at column temperatures of 220°, 160°, and 150°C, for the absolute retention time of parathion which serves as a reference figure for the calculation of the relative retention times. These values were determined on different days under the same conditions that can be realized in common laboratory practice. Because of those variations that may have different causes the relative retention times can be calculated only by comparing the glc retentions of test and reference substance, measured on the same day under the same conditions.

The electron-capture detector was also used for the glc of nitrophenol ethers and esters, for instance, in the analysis of fluorodifen (87) (WALTER *et al.* 1968), bromofenoxim (26) which was converted for this analysis into 3,5-dibromo-4-hydroxybenzoic acid and this with diazomethane into 3,5-dibromo-4-hydroxybenzoic acid methyl ester (*Ciba-Geigy AG* 1970), and dinoseb acetate (75) and binapacryl (24) (BONELLI *et al.* 1964, *Farbwerke Hoechst AG* 1972).

For the glc of the insecticides parathion (130), parathion methyl (132), fenitrothion (81), and the oxidation products of 130 and 132, paraoxon (131) and paraoxon methyl (133) (RUZICKA *et al.* 1967), thermionic detectors are especially appropriate because they are simultaneously sensitive to nitrogen and phosphorus. In some cases paraoxon (131) was observed as a main metabolite in animal organisms after administraton of parathion (130), whereas it was only temporarily detected in very small quantities in plants (lettuce, beans) after treatment with parathion (COFFIN 1966, EL-REFAI and HOPKINS 1966). Therefore, it is commonly not found as a residue in food originating from plants, as also is the case with the corresponding oxygen derivatives of other phosphorothioates.

The *N*-FID has also proved to be appropriate in the glc determination and separation of the residues of organophosphorus pesticides not containing nitrogen as well as in their multiresidue analysis, beside the detection methods used for these compounds so far, e.g., with potassium chloride or caesium bromide thermionic detectors (HARTMANN 1966, WESSEL 1967, WATTS and STORHERR 1969, KRAUSE and KIRCHHOFF 1970, RENVALL and ÅKERBLOM 1971, *Deutsche Forschungsgemeinschaft* 1972, and others), or with electron-capture detectors (BURKE and HOLSWADE 1966, WESSEL 1967, *Deutsche Forschungsgemeinschaft* 1972, and others), or with the COULSON electrolytic conductivity detector in the nitrogen mode (LASKI and WATTS 1973). Table IX shows, as examples, the retention times of the nonnitrogen but phosphorus-containing pesticides malathion (162), mevinphos (163), and fenthion (164) detected under the given operating conditions.

f) Heterocyclic compounds with one, two, and three ring-nitrogen atoms

The non-phosphorus containing, but nitrogen containing, insecticides dimetilan (73) and fenazaflor (80) mentioned in Tables X,

Table IX. *Retention times of malathion, mevinphos, and fenthion.*

No. in Table XX	Compound	Column temp.[b] (°C)	Retention times			
			Absolute (10^{-2} min)		Relative to parathion = 1.00	
			SE 30[c]	QF1[d]	SE 30[c]	QF 1[d]
162	malathion	220	190	231	0.91	0.75
		160	1,713	2,418	1.01	0.78
163	mevinphos	220	54	81	0.26	0.26
		160	188	409	0.10	0.13
164	fenthion	220	203	173	0.99	0.56
		160	1,660	1,460	0.98	0.48
		150	3,680	2,342	0.98	0.50

[b-d] See footnotes to Table III.

Table X. *Retention times of heterocyclic compounds with one ring-nitrogen of structures:*

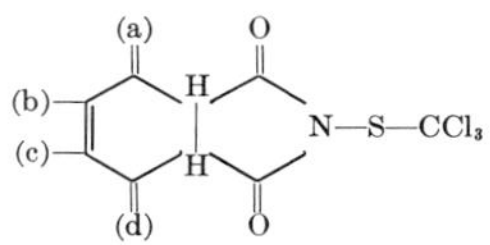

A

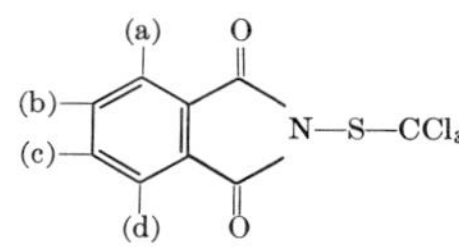

B

phthalimide-

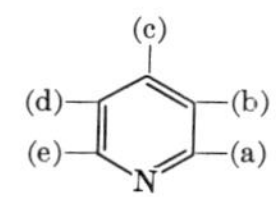

C
pyridine-

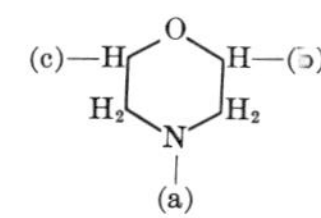

D
morpholine-

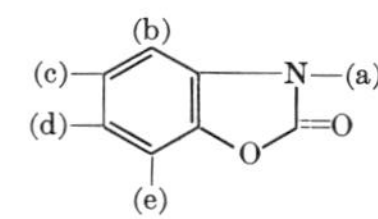

E
benzoxazol-

No. in Table XX	Structure	Substituent in the structure					Compound[a]	Column temp.[b] (°C)	Retention times			
									Absolute (10^{-2} min)		Relative to parathion = 1.00	
		(a)	(b)	(c)	(d)	(e)			SE 30[c]	QF 1[d]	SE 30[c]	QF 1[d]
31	A	H_2	H	H	H_2	—	captan	220 160	247 —	96, 334 464, 3,179	1.19 —	0.31, 1.08 0.14, 0.99
89	B	H	H	H	H	—	folpet	220 160	264 209, 2,0[illegible]0	86, 310 —	1.28 0.13, 1.24	0.28, 1.00 —
79	C	$—O—P(S)(OC_2H_5)_2$	Cl	H	Cl	Cl	dursban (chlorpyrifos)	220 160	210 1,763	157 1,214	1.01 1.07	0.51 0.39
161	D	$—CH_2—C_{12}H_{25}$	$—CH_3$	$CH_3—$	—	—	tridemorph	220 150	155 2,758	— —	0.75 0.73	— —
141	E	$—CH_2—S—P(S)(OC_2H_5)_2$	H	H	Cl	H	phosalone	220	917	1,262	4.43	4.15

[a–d] See footnotes to Table III.

XI, and XII are pyrazol (benzimidazol) derivatives. The phosphorus- and nitrogen-containing insecticides dursban (79), phosalone (141), methidathion (102) with its metabolites 103 and 104 (ESSER *et al.* 1967), and diazinon (60) with its metabolites 61 and 62 (EBERLE and NOVAK 1969), as well as diethquinalphion (68), its oxoderivative 69 (SCHMIDT and HAMMANN 1969), and azinphosethyl (16), azinphos-methyl (18) and their oxygen analogs 17 and 19 (MILES 1964), are pyridine-, benzoxazol-, thiadiazol-, pyrimidine-, quinoxaline- and 1,2,3-benzotriazine-derivatives. In glc they can be determined well with the *N*-FID. As is known, the nitrogen-containing phosphorus compounds 79, 141, 102, 103, 60, 61, 68, 69, 16, 17, 18, and 19 can also be determined with other phosphorus-sensitive alkali salt thermionic detectors (IVES and GIUFFRIDA 1967, RUZICKA *et al.* 1967, THORNTON and ANDERSON 1968, DRÄGER 1969, KRAUSE and KIRCHHOFF 1970, EBERLE and HÖRMANN 1971, RENVALL and ÅKERBLOM 1971, STORHERR *et al.* 1971, *Deutsche Forschungsgemeinschaft* 1972, and others). Phosalone (141), diazinon (60), and azinphosethyl and -methyl (16 and 18) were also analytically determined with the electron- capture detector (BURKE and HOLSWADE 1966, GUARDIGLI *et al.* 1971), and methidathion (102) and diazinon (60) with the COULSON electrolytic conductivity detector (EBERLE and HÖRMANN 1971, LASKI and WATTS 1973).

The fungicides captan (31), folpet (89), tridemorph (161), fuberidazole (91), quinomethionate (34), and anilazine (8) mentioned in Tables X, XI and XII are phthalimide-, morpholine-, benzimidazol-, quinoxaline-, and 1,3,5-triazine-derivatives. In their glc determination the *N*-FID renders as good a service as in the glc analysis of the herbicides listed in Tables XI and XII azolamid (21), norflurazon (128) and its demethylation product 129, pyrazon (151), bromacil (25), lenacil (93) on imidazol-, pyridazine- and pyrimidine-bases, as well as of the herbicide metribuzin (115) with its metabolites 116, 117, 118, which are derived from 1,2,4-triazine, and of the 1,3,5-triazine herbicides simazine (154), atrazine (13), terbuthylazine (155), propazine (148), atratone (9), prometone (145), desmetryne (57), ametryne (3), terbutryne (157), prometryne (146), methoprotryne (105), and aziprotryne (20) and their metabolites 14, 15, 156, 10, 11, 12, 58, 4, 5, 6, 158, and 106. Further details of these metabolites are presented by KNUESLI *et al.* (1969), KAUFMAN and KEARNY (1970), and JORDAN *et al.* (1970).

MATTSON *et al.* (1970) reviewed the different methods for the glc determination of the triazine herbicides. TINDLE *et al.* (1968) referred to a thermionic detector with rubidium sulphate for the glc analysis of *s*-triazine residues. According to the literature, captan (31) (ANONYMOUS 1971), dursban (79), diazinon (60), azinphosmethyl (18), simazine (154), atrazine (13), propazine (148) (WRIGHT and RINER, 1969, *Deutsche Forschungs-Gemeinschaft* 1972), lenacil (93)

Table XI. *Retention times of heterocyclic compounds with two ring-nitrogens of structures:*

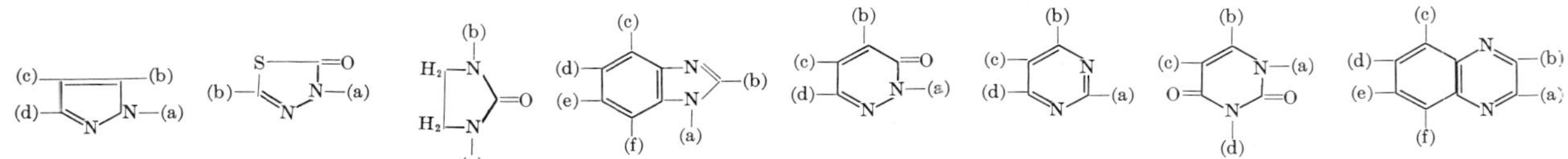

A pyrazol- B thiadizol- C imidazol- D benzimidazol- E pyridazine- F pyrimidine- G H quinoxaline-

No. in Table XX	Structure	Substituents in the structure (a)	(b)	(c)	(d)	(e)	(f)	Compound[a]	Column temp.[b] (°C)	Retention times: Absolute (10^{-2} min) SE 30[c]	QF 1[d]	Relative to parathion = 1.00 SE 30[c]	QF 1[d]
73	A	—CO—N(CH_3)$_2$	—CH_3	H	(CH_3)$_2$N—C(O)—O—	—	—	dimetilan	220 160	154 1,157	244 2,567	0.74 0.70	0.79 0.84
102	E	—CH_2—S—P(S)(OCH_3)$_2$	CH_3O—	—	—	—	—	methidathion	220 160	277 2,373	341 3,716	1.33 1.45	1.12 1.15
103	B	—CH_2—S—P(O)(OCH_3)$_2$	CH_3O—	—	—	—	—	methidathion-met	220 160	214 1,838	422 4,911	1.03 1.08	1.39 1.52
104	B	H	CH_3O—	—	—	—	—	methidathion-met	160 150	101 223	112 —	0.06 0.06	0.04 —
21	C	—CO—NH—CH_2—CH(CH_3)$_2$	H	—	—	—	—	azolamid	220 160	103 718	— —	0.55 0.49	— —
80	D	—C(O)—O—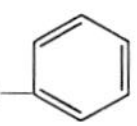	—CF_3	H	Cl	Cl	H	fenazaflor	220 160	— —	270 5,324	— —	0.87 1.65
91	D	H	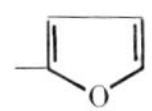	H	H	H	H	fuberidazole	220 160	161 1,110	187 1,486	0.77 0.65	0.60 0.48
128	E	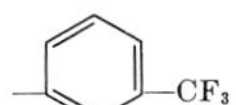	Cl	CH_3—NH—	H	—	—	norflurazon	220	514	1,226	2.48	4.03
129	E	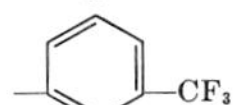	Cl	H_2N—	H	—	—	norflurazon-met	220	451	998	2.18	3.28
151	E	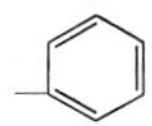	Cl	H_2N—	H	—	—	pyrazon	220	504	1,124	2.46	3.70

60	F	$—CH(CH_3)_2$	$—CH_3$	H	$(C_2H_5O)_2P(S)—O—$	—	—	diazinon	220 160	123 822	87 523	0.58 0.50	0.28 0.17
61	F	$—CH(CH_3)_2$	$—CH_3$	H	$(C_2H_5O)_2P(O)—O—$	—	—	diazinon-met	220 160	111 727	128 1,057	0.52 0.44	0.41 0.35
62	F	$—CH(CH_3)_2$	$—CH_3$	H	HO—	—	—	diazinon-met	220 160	— 256	75 362	— 0.16	0.24 0.12
25	G	H	$—CH_3$	Br	$—CH(C_2H_5)—CH_3$	—	—	bromacil	220 160	178 1,518	299 —	0.84 0.92	0.97 —
93	G	H	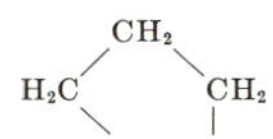		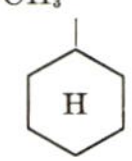	—	—	lenacil	220	520, 1,056	635, 737	2.49, 5.05	2.09, 2.42
34	H	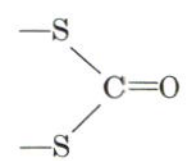		H	$CH_3—$	H	H	quinomethio-nate	220 160	289 —	219 1,623	1.36 —	0.70 0.53
68	H	$—O—P(S)(OC_2H_5)_2$	H	H	H	H	H	diethquin-alphion	220 160	261 2,386	227 2,155	1.27 1.45	0.73 0.71
69	H	$—O—P(O)(OC_2H_5)_2$	H	H	H	H	H	diethquin-alphion-met	220 160	233 2,107	369 4,396	1.13 1.28	1.21 1.36

[a–d] See footnotes to Table III.

Table XII. *Retention times of heterocyclic compounds with three ring-nitrogens of structures:*

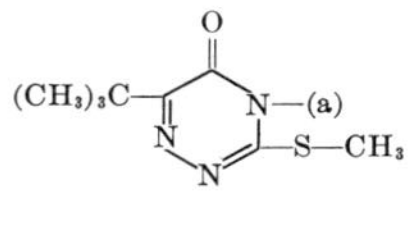

A

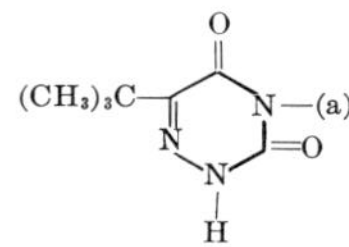

B

1,2,4-triazine-

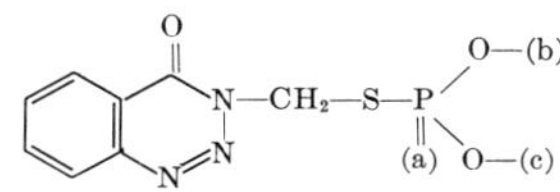

C

1,2,3-benzotriazine-

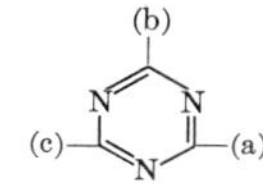

D

1,3,5-triazine-

No. in Table XX	Structure	Substituents in the structure			Compound[a]	Column temp.[b] (°C)	Retention times			
							Absolute (10^{-2} min)		Relative to parathion = 1.00	
		(a)	(b)	(c)			SE 30[c]	QF 1[d]	SE 30[c]	QF 1[d]
115	A	—NH_2	—	—	metribuzin	220 160	139 1,108	— —	0.74 0.76	— —
116	B	—NH_2	—	—	metribuzin-met	220 160	85 465	— —	0.46 0.32	— —
117	A	H	—	—	metribuzin-met	220 160	126 1,019	— —	0.67 0.69	— —
118	B	H	—	—	metribuzin-met	220 160	62, 345 293	— —	0.33, 1.85 0.20	— —
16	C	S	—C_2H_5	—	azinphosethyl	220	1,156	1,416	5.43	4.57
17	C	O	—C_2H_5	—	azinphosethyl-met	220	895	1,841	4.20	5.94
18	C	S	—CH_3	—	azinphosmethyl	220	903	1,222	4.24	3.94
19	C	O	—CH_3	—	azinphosmethyl-met	220	682	—	3.20	—
154	D	Cl	—NH—C_2H_5	C_2H_5—NH—	simazine	220 160	104 577	99 579	0.50 0.35	0.32 0.18
13	D	Cl	—NH—C_2H_5	$(CH_3)_2$CH—NH—	atrazine	220 160	110 613	99 602	0.52 0.37	0.32 0.20
14	D	Cl	—NH—C_2H_5	H_2N—	atrazine-met	220 160	88 408	85 427	0.41 0.25	0.28 0.14
15	D	Cl	—NH_2	$(CH_3)_2$CH—NH—	atrazine-met	220 160	90 432	86 441	0.42 0.26	0.28 0.14
155	D	Cl	—NH—C_2H_5	$(CH_3)_3$C—NH—	terbuthylazine	220 160	116 712	107 687	0.56 0.43	0.35 0.22
156	D	Cl	—NH_2	$(CH_3)_3$C—NH—	terbuthylazine-met	220 160	93 484	90 478	0.45 0.29	0.29 0.15
148	D	Cl	—NH—CH$(CH_3)_2$	$(CH_3)_2$CH—NH—	propazine	220 160	109 649	100 615	0.53 0.40	0.32 0.20
8	D	Cl	Cl	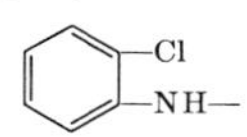	anilazine	220 160	256 2,046	183 1,463	1.20 1.24	0.59 0.48

9	D	—OCH_3	—NH—C_2H_5	$(CH_3)_2$CH—NH—	atratone	220	—	88	—	0.28
						160	582	535	0.35	0.17
10	D	—OCH_3	—NH—C_2H_5	H_2N—	atratone-met	220	82	81	0.39	0.26
						160	371	425	0.22	0.14
11	D	—OCH_3	—NH_2	$(CH_3)_2$CH—NH—	atratone-met	220	83	82	0.40	0.27
						160	397	435	0.24	0.14
12	D	—OCH_3	—NH_2	H_2N—	atratone-met	220	65	70	0.32	0.23
						160	259	326	0.16	0.10
145	D	—OCH_3	—NH—CH$(CH_3)_2$	$(CH_3)_2$CH—NH—	prometone	220	105	88	0.51	0.28
						160	619	536	0.38	0.18
57	D	—SCH_3	—NH—CH_3	$(CH_3)_2$CH—NH—	desmetryne	220	147	122	0.69	0.39
						160	1,001	859	0.61	0.28
58	D	—SCH_3	—NH—CH_3	H_2N—	desmetryne-met	220	113	107	0.53	0.35
						160	631	648	0.38	0.21
3	D	—SCH_3	—NH—C_2H_5	$(CH_3)_2$CH—NH—	ametryne	220	165	129	0.78	0.42
						160	1,200	1,272	0.73	0.42
4	D	—SCH_3	—NH—C_2H_5	H_2N—	ametryne-met	220	128	115	0.60	0.37
						160	773	755	0.47	0.25
5	D	—SCH_3	—NH_2	$(CH_3)_2$CH—NH—	ametryne-met	220	132	115	0.62	0.37
						160	826	771	0.50	0.25
6	D	—SCH_3	—NH_2	H_2N—	ametryne-met	220	—	99	—	0.32
						160	523	587	0.32	0.19
157	D	—SCH_3	—NH—C_2H_5	$(CH_3)_3$C—NH—	terbutryne	220	179	137	0.86	0.44
						160	1,402	1,092	0.85	0.35
158	D	—SCH_3	—NH_2	$(CH_3)_3$C—NH—	terbutryne-met	220	141	121	0.68	0.39
						160	921	841	0.56	0.27
146	D	—SCH_3	—NH—CH$(CH_3)_2$	$(CH_3)_2$CH—NH—	prometryne	220	165	127	0.80	0.41
						160	1,244	972	0.76	0.31
105	D	—SCH_3	—NH—CH$(CH_3)_2$	CH_3O—$(CH_2)_3$—NH—	methoprotryne	220	363	280	1.72	0.90
						160	—	3,340	—	1.08
106	D	—SCH_3	—NH_2	CH_3O—$(CH_2)_3$—NH—	methoprotryne-met	220	283	255	1.35	0.82
						160	2,829	2,661	1.72	0.86
20	D	—SCH_3	N_3	$(CH_3)_2$CH—NH—	aziprotryne	220	134	117	0.63	0.38
						160	792	788	0.48	0.26

[a–d] See footnotes to Table III.

(PEASE 1966), for instance, can be detected with flame ionization detectors. The COULSON electrolytic conductivity detector can be used for the analysis of captan (31), folpet (89), quinomethionate (34), anilazine (8), atrazine (13), propazine (148), prometone (145), ametryne (3), terbutryne (157), and prometryne (146) (LASKI and WATTS 1973). Microcoulometric detectors suit the glc determination of captan (31), folpet (89), anilazine (8), simazine (154), atrazine (13), propazine (148), desmetryne (57), terbutryne (157), prometryne (146), methoprotryne (105) (BURKE and HOLSWADE 1966, *Deutsche Forschungs-Gemeinschaft* 1972), and methidathion (102) (MATTSON *et al.* 1969). With electron-capture detectors captan (31), folpet 89, anilazine (8), simazine (154), atrazine (13), propazine (148) (BURKE and GIUFFRIDA 1964, BURKE and HOLSWADE 1966, POMERANTZ and ROSS 1968, WINDHAM 1969, MENDOZA and SHIELDS 1971), quinomethionate (34) (VOGELER and NIESSEN 1967), and bromacil (25) (JOLLIFFE *et al.* 1967, BEVENUE and OGATA 1970) were detected by glc.

IV. Conclusions

Two qualities are inherent to the *N*-FID: the detector selectivity and the sensitivity. Detector selectivity is required in any area of pesticide analysis mentioned in this review. Selectivity allows in many cases use of simple cleanup procedures or makes cleanup unnecessary in residue analysis and metabolite studies. Selectivity makes the detector usable in a wide range of applications in pesticide analysis.

The low levels of pesticide and metabolite residue tolerances in food, feed, water, etc. require sensitivity in the range of pg quantities and sub-ppb concentrations. This sensitivity is met by the 15161 B *N*-FID.

It is suggested that 4-mm id columns be used in all cases where cleanup of the sample is not required or has been simplified due to higher sample loading on the column. At the same time this increases the lowest detectable concentration.

The *N*-FID is easily adjusted to the optimum operating conditions with respect to the required selectivity and sensitivity by the mechanical means provided. Symmetrical, narrow peaks are produced by the high detector block temperature required for high sensitivity and selectivity operations. Further, the formation of deposits of sample material encountered for detector designs showing cold spots is avoided, allowing the detector to be operational over long periods of time before any cleaning is required. At normal operating conditions the detector crystal was found to be workable over several years. The crystal is easily exchangeable for a crystal with different element selectivities.

The linear dynamic range of 10^5 for the *N*-FID covers by far the

Table XIII. *Retention times relative to parathion = 1.00* (6 ft 3 mm id glass column packed with 5% SE 30 on 80/100 mesh Chromosorb W (HP), column temperature 220°C).

RRT[a]	Compound[b]
0.25	dichlobenil (63), dimefox (70)
0.26	mevinphos (163)
0.32	atratone-met (12), carbaryl (32), fluorodifen-met (88)
0.33	metribuzin-met (118)
0.34	chlorbromuron-met (37)
0.35	chlorbromuron-met (36), mercaptodimethur (98)
0.36	fluometuron-met (86)
0.38	dimethoate-met (omethoate) (72)
0.39	atratone-met (10)
0.40	atratone-met (11)
0.41	atrazine-met (14), propachlor (147)
0.42	atrazine-met (15), chlorpropham (49), dichlofluanid-met (66), dicrotophos (67)
0.43	cycloate (55), dichlobenil-met (64), DNOC (78), monocrotophos (120)
0.44	chlordimeform (40)
0.45	methabenzthiazuron (101), terbuthylazine-met (156)
0.46	metribuzin-met (116)
0.47	phosphamidon-met (143)
0.50	chlorbufam (39), simazine (154)
0.51	dimethoate (71), prometone (145)
0.52	atrazine (13), diazinon-met (61)
0.53	desmetryne-met (58), propazine (148)
0.55	aminocarb (7), azolamid (21), monolinuron (121)
0.56	terbuthylazine (155)
0.58	cycluron (56), diazinon (60)
0.59	phosphamidon-met (143)
0.60	ametryne-met (4), phosphamidon (142)
0.61	parathion methyl-met (133)
0.62	ametryne-met (5), quintozene (152)
0.63	aziprotryne (20)
0.67	dinoseb (74), formothion (90), metribuzin-met (117)
0.68	terbutryne-met (158)
0.69	desmetryne (57), tri-allate (160)
0.71	metobromuron (107), monalide (119), phosphamidon (142)
0.72	acephate (1)
0.73	phenmedipham-met (138)
0.74	dimetilan (73), metribuzin (115)
0.75	tridemorph (161)
0.76	quintozene-met (153)
0.77	fuberidazole (91), mercaptodimethur-met (99), parathion methyl (132)
0.78	ametryne (3)
0.80	parathion-met (131), prometryne (146)
0.82	mercaptodimethur-met (100)
0.84	bromacil (25)
0.85	dinoseb acetate (75)
0.86	chloroxuron (46), terbutryne (157)
0.87	phenmedipham (137)
0.89	fenitrothion (81)
0.91	dichlofluanid (65), malathion (162)
0.92	chloroxuron-met (48)
0.93	chloroxuron-met (47), linuron (94)
0.99	fenthion (164)

Table XIII. (*Continued*)

RRT[a]	Compound[b]
1.00	parathion (130)
1.01	dursban (79)
1.03	methidathion-met (103)
1.10	monolinuron-met (125)
1.13	diethquinalphion-met (69)
1.19	captan (31)
1.20	anilazine (8)
1.22	chlorbromuron (35)
1.24	chloroxuron-met (48), phenmedipham-met (140)
1.27	diethquinalphion (68)
1.28	folpet (89)
1.33	methidathion (102)
1.35	methoprotryne-met (106)
1.36	quinomethionate (34)
1.48	buturon (28)
1.51	fluorodifen (87)
1.52	phenamiphos (134)
1.54	barban (22)
1.72	methoprotryne (105)
1.77	chlorfensulphide (45)
1.82	chlordimeform-met (44)
1.85	metribuzin-met (118)
1.91	binapacryl (24)
1.99	metobromuron-met (111)
2.18	norflurazon-met (129)
2.23	phenmedipham-met (139)
2.28	chlorfensulphide (45), chlortoluron-met (54)
2.46	pyrazon (151)
2.48	norflurazon (128)
2.49	lenacil (93)
3.17	nitralin (127)
3.20	azinphosmethyl-met (19)
3.31	phenamiphos-met (135)
3.35	phenamiphos-met (136)
3.36	linuron-met (97)
4.20	azinphosethyl-met (17)
4.24	azinphosmethyl (18)
4.43	phosalone (141)
5.05	lenacil (93)
5.10	linuron-met (97)
5.23	benzthiazuron (23)
5.43	azinphosethyl (16)
6.25	chlorbromuron-met (38)

[a] RRT = relative retention time.

[b] The numbers in parentheses are the numbers under which the compounds have been tested. "Met" indicates a metabolite, degradation product, or impurity in the corresponding pesticide. Their chemical constitutions are listed in tables III, IV, V, VII, VIII, X, XI, and XII.

Table XIV. *Retention times relative to parathion = 1.00* (6 ft 2 mm id glass column packed with 5% SE 30 on 80/100 mesh Chromosorb W (HP), column temperature 160°C).

RRT[a]	Compound[b]
0.05	acephate-met (methamidophos) (2), monolinuron-met (124)
0.06	dimefox (70), methidathion-met (104)
0.07	chlordimeform-met (43), chlortoluron-met (53), metobromuron-met (110), neburon (126)
0.08	chlorthiamid (50), dichlobenil (63), diuron (77), linuron-met (95), metobromuron-met (109), monolinuron-met (124)
0.10	butoxicarboxim (29), chlortoluron-met (53), dioxacarb (76), metabromuron-met (110), metoxuron-met (113), mevinphos (163)
0.11	linuron-met (96), metoxuron (112), metoxuron-met (114)
0.12	carbaryl (32), propham (149)
0.13	folpet (89)
0.14	acephate (1), linuron-met (96)
0.15	fluorodifen-met (88)
0.16	atratone-met (12), butylate (30), chlorbufam (39), diazinon-met (62)
0.17	metoxuron-met (114)
0.18	mercaptodimethur (98)
0.19	chlordimeform-met (42)
0.20	fluometuron-met (86), metribuzin-met (118)
0.22	atratone-met (10), DNOC (78), propachlor (147), propoxur (150)
0.24	atratone-met (11)
0.25	atrazine-met (14), cycloate (55), dichlofluanid-met (66), fluometuron (82)
0.26	atrazine-met (15), chlorpropham (49)
0.27	chlordimeform-met (41), dichlobenil-met (64), methabenzthiazuron (101)
0.28	monocrotophos (120)
0.29	terbuthylazine-met (156)
0.32	ametryne-met (6), metribuzin-met (116)
0.33	dimethoate (71)
0.34	di-allate (59), fluometuron-met (83)
0.35	atratone (9), fluometuron-met (84), simazine (154)
0.37	atrazine (13)
0.38	chlordimeform (40), desmetryne-met (58), prometone (145)
0.40	propazine (148)
0.41	monolinuron (121), phosphamidon-met (143)
0.42	cycluron (56), quintozene (152)
0.43	terbuthylazine (155)
0.44	aminocarb (7), diazinon-met (61)
0.47	ametryne-met (4), parathion methyl-met (133)
0.48	aziprotryne (20)
0.49	azolamide (21), phosphamidon (142), phosphamidon-met (143)
0.50	ametryne-met (5), diazinon (60), dinoseb (74), monolinuron-met (123)
0.53	formothion (90)
0.55	tri-allate (160)
0.56	quintozene-met (153), terbutryne-met (158)
0.58	monalide (119)
0.59	dioxacarb (76), metobromuron (107)
0.61	desmetryne (57)
0.64	parathion methyl (132), phosphamidon (142)
0.65	fuberidazole (91)
0.66	phenmedipham-met (138)
0.69	metribuzin-met (117)
0.70	dimetilan (73)
0.73	ametryne (3)

Table XIV. (*Continued*)

RRT[a]	Compound[b]
0.75	mercaptodimethur-met (99), monolinuron-met (122)
0.76	metribuzin (115), prometryne (146)
0.77	chloroxuron (46)
0.80	dinoseb acetate (75)
0.81	mercaptodimethur-met (100), parathion-met (131)
0.82	fenitrothion (81)
0.84	chloroxuron-met (48)
0.85	mercaptodimethur (98), terbutryne (157)
0.86	dichlofluanid (65)
0.88	linuron (94)
0.92	bromacil (25)
0.98	fenthion (164)
1.00	parathion (130)
1.01	bromofenoxim (26), malathion (162)
1.07	dursban (79), monolinuron-met (125)
1.08	methidathion-met (103)
1.16	metobromuron-met (108)
1.24	anilazine (8), folpet (89)
1.28	diethquinalphion-met (69)
1.29	chlorbromuron (35)
1.36	chloroxuron-met (48)
1.44	phenmedipham-met (140)
1.45	diethquinalphion (68), methidathion (102)
1.72	methoprotryne-met (106)
2.17	barban (22)

[a-b] See footnotes to Table XIII.

concentration range encountered in the many areas of pesticide analysis. The COULSON detector is superior by a factor of about 10 to the *N*-FID with respect to selectivity, while 10^2 to 10^3 times less sensitive. The larger sample volumes injected on the column using the COULSON detector to achieve the required sensitivity, of course, in practice do not result in any selectivity advantages. While very sensitive to polyhalogenated compounds, the electron-capture detector is not sensitive enough to nitrogen-containing pesticides, except for compounds containing the nitro group. Efficient cleanup procedures are through the way required for the electron-capture detector. The flame photometric detector is slightly more selective, but about 50 times less sensitive, than the *N*-FID in pesticide analysis of phosphorus-containing compounds. The highest selectivity among all the detectors coupled to glc is achieved by mass spectrometry requiring effective cleanup steps. The only selective detector sensitive to pg quantities of all gas chromatographically eluted nitrogen- and phosphorus-containing pesticides, generally applicable to the various areas of pesticide analysis, is the *N*-FID.

The 164 compounds evaluated are arranged in Tables XIII to XVIIII according to their relative retention times on the SE 30 column and on the QF 1 column, at column temperatures of 220°, 160°, and 150°C. We chose for the evaluation these two columns because of their differing polarity and their good selectivity that has proved to be appropriate in many laboratories in glc pesticide analysis. Other column packings known through the literature would have submitted a better selectivity with some compounds, but we did not use them for we wanted to achieve comparable testing results.

It is not possible to expect all compounds mentioned here to pass the glc columns unchanged or controllably changed. Therefore, the determined retention times may indicate the pesticides brought into the tests, their metabolites, or contaminations, as well as secondary products formed during the analysis. The extent of the formation and the glc behaviour of such secondary products under different conditions are essential for the question whether a substance positively can be detected with glc methods. Such secondary products may

Table XV. *Retention times relative to parathion = 1.00*
(6 ft 2 mm id glass column packed with 5% SE 30 on 80/100 mesh Chromosorb W (HP), column temperature 150°C).

RRT[a]	Compound[b]
0.02	fluometuron (82)
0.03	fluometuron-met (85), thiophanate (159)
0.04	phoxim (144)
0.05	cartap (33), chlortoluron (51)
0.06	chlordimeform-met (43), chlortoluron-met (52, 53), methidathion-met (104), metobromuron-met (110), phoxim (144)
0.07	diuron (77), linuron-met (95), metobromuron-met (109), neburon (126)
0.08	chlordimeform-met (43), chlorthiamid (50)
0.09	chlortoluron-met (53), metobromuron-met (110), metoxuron (112)
0.10	linuron-met (96), metoxuron-met (113, 114)
0.11	butylate (30), propham (149)
0.13	linuron-met (96)
0.15	chlorbromuron-met (37)
0.16	metoxuron-met (114)
0.19	fluometuron-met (86)
0.20	dimethoate-met (omethoate) (72)
0.24	cycloate (55)
0.31	di-allate (59)
0.32	di-allate (59)
0.48	phosphamidon (142)
0.51	phosphamidon-met (143)
0.63	phosphamidon (142)
0.73	tridemorph (161)
0.98	fenthion (164)
1.00	parathion (130)

[a–b] See footnotes to Table XIII.

Table XVI. *Retention times relative to parathion = 1.00*
(6 ft 2 mm id glass column packed with 5% QF 1 on 80/100 mesh Chromosorb W (HP), column temperature 220°C).

RRT[a]	Compound[b]
0.20	carbaryl (32), dichlobenil (63)
0.21	cycloate (55), fluometuron-met (86)
0.22	chlorbromuron-met (37)
0.23	atratone-met (12), chlorbromuron-met (36)
0.24	di-allate (59), diazinon-met (62)
0.26	atratone-met (10), mevinphos (163)
0.27	atratone-met (11), chlordimeform (40), chlordimeform-met (41), chlorpropham (49)
0.28	atratone (9), atrazine-met (14, 15), diazinon (60), fluorodifen-met (88), folpet (89), methabenzthiazuron (101), prometone (145), tri-allate (160)
0.29	chlorbufam (39), terbuthylazine-met (156)
0.31	captan (31), phenmedipham-met (138)
0.32	ametryne-met (6), atrazine (13), propazine (148), simazine (154)
0.35	bromofenoxim-met (27), desmetryne-met (58), propachlor (147), quintozene (152), terbuthylazine (155)
0.36	quintozene-met (153)
0.37	ametryne-met (4, 5), chlordimeform-met (42)
0.38	aziprotryne (20)
0.39	desmetryne (57), terbutryne-met (158)
0.40	aminocarb (7)
0.41	diazinon-met (61), prometryne (146)
0.42	ametryne (3)
0.44	chloroxuron (46), DNOC (78), terbutryne (157)
0.46	cycluron (56), dichlobenil-met (64), monolinuron (121), phenmedipham (137)
0.47	chloroxuron-met (48)
0.48	chloroxuron-met (47), fluometuron-met (83)
0.49	monalide (119)
0.51	dichlofluanid-met (66), dursban (79)
0.52	fluometuron-met (84)
0.55	dimethoate-met (omethoate) (72)
0.56	fenthion (164)
0.58	metobromuron (107)
0.59	anilazine (8), chloroxuron-met (48), dimethoate (71)
0.60	fuberidazole (91), mercaptodimethur (98), phenmedipham-met (140)
0.66	dicrotophos (67), dinoseb (74)
0.67	chlordimeform-met (44), chlorfensulphide (45)
0.68	monocrotophos (120)
0.69	ioxynil (92)
0.70	metobromuron-met (111), quinomethionate (34)
0.73	diethquinalphion (68)
0.75	malathion (162)
0.79	dimetilan (73), phosphamidon-met (143)
0.80	linuron (94)
0.81	formothion (90)
0.82	chlorfensulphide (45), chlortoluron-met (54), methoprotryne-met (106), parathion methyl (132)
0.83	dichlofluanid (65)
0.87	fenazaflor (80), fenitrothion (81)
0.88	phosphamidon (142)
0.89	barban (22)

Table XVI (*Continued*)

RRT[a]	Compound[b]
0.90	methoprotryne (105)
0.96	mercaptodimethur-met (99)
0.97	bromacil (25)
1.00	folpet (89), parathion (130)
1.03	parathion methyl-met (133)
1.07	mercaptodimethur-met (100)
1.08	captan (31)
1.12	methidathion (102)
1.15	linuron-met (97)
1.21	diethquinalphion-met (69), dinoseb acetate (75)
1.24	phosphamidon (142)
1.28	parathion-met (131)
1.29	phenmedipham-met (139)
1.36	phenamiphos (134)
1.39	methidathion-met (103)
1.74	buturon (28)
1.95	chlorbromuron-met (38)
2.06	linuron-met (97)
2.08	benzthiazuron (23)
2.09	lenacil (93)
2.33	binapacryl (24)
2.42	lenacil (93)
3.27	fluorodifen (87)
3.28	norflurazon-met (129)
3.70	pyrazon (151)
3.94	azinphosmethyl (18)
4.03	norflurazon (128)
4.15	phosalone (141)
4.57	azinphosethyl (16)
5.94	azinphosethyl-met (17)
7.70	nitralin (127)
9.03	phenamiphos-met (135)
10.17	phenamiphos-met (136)

[a–b] See footnotes to Table XIII.

especially disturb the multiresidue analysis. In some cases it is recommended to form appropriate derivatives.

Two or more chemically different compounds often show on the same glc columns and under the same temperature conditions the same or similar retention times. Corresponding examples are given in Table XIX in the case of aziprotryne-phosphamidon, diazinon-dinoseb, and bromofenoxim-malathion on the SE 30 column, and in the case of metobromuron-anilazine, fenthion-fuberidazole, dichlofluanid-formothion, and atrazine-propazine on the QF 1 column. In glc analysis the risk of mixing up different compounds because of the same or similar retention times does not exist when the retention times of these substances obviously differ on other available column

Table XVII. *Retention times relative to parathion = 1.00* (6 ft 2 mm id glass column packed with 5% QF 1 on 80/100 mesh Chromosorb W (HP), column temperature 160°C).

RRT[a]	Compound[b]
0.02	fluometuron-met (85)
0.03	chlortoluron-met (53), diuron (77), linuron-met (95), monolinuron-met (124), neburon (126)
0.04	butylate (30), chlordimeform-met (43), chlortoluron-met (53), methidathion-met (104), metobromuron-met (109, 100), monolinuron-met (124)
0.05	metobromuron-met (110), metoxuron-met (113)
0.06	linuron-met (96), mercaptodimethur (98), metoxuron (112), propham (149)
0.07	carbaryl (32), chlorthiamid (50), dichlobenil (63), dioxacarb (76), metoxuron-met (114)
0.08	chlorbromuron-met (37), cycloate (55)
0.09	chlorbromuron-met (36), fluometuron-met (86)
0.10	atratone-met (12), di-allate (59), metoxuron-met (114)
0.12	di-allate (59), diazinon-met (62), fluorodifen-met (88)
0.13	chlordimeform (40), chlordimeform-met (41), methabenzthiazuron (101), mevinphos (163)
0.14	atratone-met (10, 11), atrazine-met (14, 15), captan (31), chlorpropham (49)
0.15	terbuthylazine-met (156), tri-allate (160)
0.17	atratone (9), chlorbufam (39), diazinon (60)
0.18	bromofenoxim-met (27), phenmedipham-met (138), prometone (145), quintozene (152), simazine (154)
0.19	ametryne-met (6), propoxur (150), quintozene-met (153)
0.20	atrazine (13), propachlor (147), propazine (148)
0.21	bromofenoxim (26), desmetryne-met (58)
0.22	chlordimeform-met (42), terbuthylazine (155)
0.25	ametryne-met (4, 5)
0.26	aziprotryne (20), DNOC (78)
0.27	terbutryne-met (158)
0.28	desmetryne (57)
0.29	aminocarb (7), phenmedipham (137)
0.30	chloroxuron (46), dichlobenil-met (64)
0.31	prometryne (146)
0.33	monlinuron (121)
0.34	chloroxuron-met (48)
0.35	cycluron (56), diazinon-met (61), terbutryne (157)
0.36	dichlofluanid-met (66)
0.37	monalide (119)
0.39	dursban (79), fluometuron-met (83)
0.42	ametryne (3)
0.45	dimethoate (71)
0.47	ioxynil (92), metobromuron (107)
0.48	anilazine (8), fenthion (164), fuberidazole (91), monolinuron-met (123)
0.49	fluometuron-met (84)
0.51	dinoseb (74), chloroxuron-met (48), phenmedipham-met (140)
0.53	mercaptodimethur (98), quinomethionate (34)
0.57	chlorfensulphide (45), metobromuron-met (111)
0.59	chlordimeform-met (44)
0.62	dicrotophos (67)
0.64	monocrotophos (120)
0.70	dioxacarb (76)

Table XVII. (*Continued*)

RRT[a]	Compound[b]
0.71	diethquinalphion (68), parathion methyl (132)
0.74	linuron (94), monlinuron-met (122)
0.76	chlorfensulphide (45)
0.77	dichlofluanid (65), formothion (90)
0.78	chlortoluron-met (54), malathion (162)
0.81	phosphamidon-met (143)
0.84	dimetilan (73)
0.85	fenitrothion (81)
0.86	methoprotryne-met (106)
0.89	mercaptodimethur-met (99), phosphamidon (142)
0.97	barban (22), parathion methyl-met (133)
0.99	captan (31), mercaptodimethur-met (100)
1.00	parathion (130)
1.05	metobromuron-met (108)
1.08	methoprotryne (105)
1.15	methidathion (102)
1.20	linuron-met (97)
1.25	dinoseb acetate (75)
1.36	diethquinalphion-met (69)
1.42	parathion-met (131)
1.52	methidathion-met (103), phenmedipham-met (139)
1.65	fenazaflor (80)
1.77	phenamiphos (134)
2.02	buturon (28)

[a–b] See footnotes to Table XIII.

Table XVIII. *Retention times relative to parathion = 1.00* (6 ft 2 mm id glass column packed with 5% QF 1 on 80/100 mesh Chromosorb W (HP), column temperature 150°C).

RRT[a]	Compound[b]
0.01	fluometuron (82), fluometuron-met (85), thiophanate (159)
0.02	chlortoluron (51) chlortoluron-met (52)
0.03	chlordimeform-met (43), chlortoluron-met (53), diuron (77), metobromuron-met (110), neburon (126)
0.04	butylate (30), chlordimeform-met (43), metobromuron-met (109)
0.05	chlortoluron-met (53), metobromuron-met (110), metoxuron (112), metoxuron-met (113)
0.06	chlorthiamid (50), linuron-met (95, 96)
0.07	linuron-met (96), metoxuron-met (114), propham (149)
0.08	chlorbromuron-met (37)
0.09	chlorbromuron-met (36), fluometuron-met (86)
0.10	cycloate (55)
0.11	di-allate (59)
0.12	metoxuron-met (114)
0.13	di-allate (59)
0.28	dimethoate-met (omethoate) (72)
0.50	fenthion (164)
1.00	parathion (130)

[a–b] See footnotes to Table XIII.

Table XIX. *Comparison of relative retention times of some compounds on SE 30 and QF 1 (column temperature* 160°C).

No. in Table XX	Compound[a]	RRT		No. in Table XX	Compound[a]	RRT		No. in Table XX	Compound[a]	RRT	
		SE 30[b,c]	QF 1[b,d]			SE 30[b,c]	QF 1[b,d]			SE 30[b,c]	QF 1[b,d]
20	aziprotryne	0.48	0.26	107	metobromuron	0.59	0.47	77	diuron	0.08	0.03
142	phosphamidon	0.49	0.89	8	anilazine	1.24	0.48	126	neburon	0.07	0.03
60	diazinon	0.50	0.17	164	fenthion	0.98	0.48	15	atrazine-met	0.26	0.14
74	dinoseb	0.50	0.51	91	fuberidazole	0.65	0.48	49	chlorpropham	0.26	0.14
26	bromofenoxim	1.01	0.21	65	dichlofluanid	0.86	0.77	13	atrazine	0.37	0.20
162	malathion	1.01	0.78	90	formothion	0.53	0.77	148	propazine	0.40	0.20

[a–d] See footnotes to Table III.

Table XX. *Pesticides and their degradation products mentioned in text.*

No.	Common name	Use[a]	Degradation product of	Chemical name	Retention time in Table
1	acephate	I	—	*O,S*-dimethyl-*N*-acetyl-amido phosphorothioate	III
2	methamidophos (acephate-met)	I	1	*O,S*-dimethylamido phosphorothioate	III
3	ametryne	H	—	2-methylthio-4-ethylamino-6-isopropylamino-*s*-triazine	XII
4	—	—	3, 157	2-methylthio-4-ethylamino-6-amino-*s*-triazine	XII
5	—	—	3, 57, 105, 146	2-methylthio-4-amino-6-isopropylamino-*s*-triazine	XII
6	—	—	3, 57, 105, 146, 157	2-methylthio-4,6-diamino-*s*-triazine	XII
7	aminocarb	I	—	4-dimethylamino-3-tolyl *N*-methylcarbamate	IV
8	anilazine	F	—	2,4-dichloro-6-(*o*-chloroanilino)-*s*-triazine	XII
9	atratone	H	—	2-methoxy-4-ethylamino-6-isopropylamino-*s*-triazine	XII
10	—	—	9	2-methoxy-4-ethylamino-6-amino-*s*-triazine	XII
11	—	—	9, 145	2-methoxy-4-amino-6-isopropylamino-*s*-triazine	XII
12	—	—	9, 145	2-methoxy-4,6-diamino-*s*-triazine	XII
13	atrazine	H	—	2-chloro-4-ethylamino-6-isopropylamino-*s*-triazine	XII
14	—	—	13, 154, 155	2-chloro-4-ethylamino-6-amino-*s*-triazine	XII
15	—	—	13, 148	2-chloro-4-amino-6-isopropylamino-*s*-triazine	XII
16	azinphosethyl	I	—	*O,O*-diethyl *S*[4-oxo-1,2,3-benzotriazin-3(4H)-ylmethyl] phosphorodithioate	XII
17	—	—	16	*O,O*-diethyl *S*[4-oxo-1,2,3-benzotriazin-3(4H)-ylmethyl] phosphorothioate	XII
18	azinphosmethyl	I	—	*O,O*-dimethyl *S*[4-oxo-1,2,3-benzotriazin-3(4H)-ylmethyl] phosphorodithioate	XII
19	—	—	18	*O,O*-dimethyl *S*[4-oxo-1,2,3-benzotriazin-3(4H)-ylmethyl] phosphorothioate	XII
20	aziprotryne	H	—	2-methylthio-4-azido-6-isopropylamino-*s*-triazine	XII
21	azolamid	H	—	imidazolidine (2-on) 1-carbon isobutylamide	XI

Table XX. (*Continued*)

No.	Common name	Use[a]	Degradation product of	Chemical name	Retention time in Table
22	barban	H	—	4-chloro-2-butynyl *N*-(3-chlorophenyl) carbamate	IV
23	benzthiazuron	H	—	3-(2-benzthiazolyl)-1-methylurea	V
24	binapacryl	I	—	2-*sec*-butyl-4,6-dinitrophenyl 3,3-dimethyl-acrylate	VIII
25	bromacil	H	—	5-bromo-3-*sec*-butyl-6-methyluracil	XI
26	bromofenoxim	H	—	3,5-dibromo-4-hydroxy-benzaldoxim *O* (2,4-dinitrophenolate)	VIII
27	—	—	26	2,4-dinitrophenol	VIII
28	buturon	H	—	3-(4-chlorophenyl)-1-methyl-1-(1-methyl-2-propynyl) urea	V
29	butoxicarboxim	I	—	2-methylsulfonyl *O*(*N*-methylcarbamoyl)-butanonoxim-(3)	IV
30	butylate	H	—	*N*,*N*-diisobutyl-*S*-ethyl-thiolcarbamate	IV
31	captan	F	—	*N*-trichloromethylthio tetra-hydrophthalimide	X
32	carbaryl	I	—	1-naphthyl *N*-methylcarbamate	IV
33	cartap	I	—	1,3-bis(carbamoylthio)-2-dimethylamino propane hydrochloride	III
34	chinomethionat (quinomethionate)	F	—	6-methyl-2-oxo-1,3-dithio-(4,5-b) quinoxaline	XI
35	chlorbromuron	H	—	3-(4-bromo-3-chlorophenyl)-1-methoxy-1-methylurea	V
36	—	—	35	3-(4-bromo-3-chlorophenyl) urea	V
37	—	—	35	4-bromo-3-chloroaniline	V
38	—	—	35	4,4′-dibromo-3,3-dichloro-azobenzene	V
39	chlorbufam	H	—	butin(1)-yl(3)-*N*-(3-chlorophenyl) carbamate	IV
40	chlordimeform	I	—	*N*′-(2-methyl-4-chloro-phenyl)-*N*,*N*-dimethyl-formamidine	VII
41	—	—	40	*N*′-(2-methyl-4-chloro-phenyl)-*N*-methyl-formamidine	VII
42	—	—	40	*N*-formyl-2-methyl-4-chloroaniline	VII
43	—	—	40	2-methyl-4-chloroaniline	VII

Table XX. (*Continued*)

No.	Common name	Use[a]	Degradation product of	Chemical name	Retention time in Table
44	—	—	40	2,2′-dimethyl-4,4′-dichloro-azobenzene	VII
45	chlorfensulphide	I	—	4-chlorophenyl-(2′,4′,5′-tri-chlorophenyl-azo) sulfide	VII
46	chloroxuron	H	—	3-[4-(4-chlorophenoxy)-phenyl]-1,1-dimethylurea	V
47	—	—	46	3-[4-(4-chlorophenoxy)-phenyl]-urea	V
48	—	—	46	4-(4-chlorophenoxy)-aniline	V
49	chlorpropham	H	—	isopropyl *N*-(3-chlorophenyl) carbamate	IV
50	chlorthiamid	H	—	2,6-dichlorothiobenzamide	III
51	chlortoluron	H	—	3-(3-chloro-4-methylphenyl)-1,1-dimethylurea	V
52	—	—	51	3-(3-chloro-4-methylphenyl)-1-methylurea	V
53	—	—	51	3-chloro-4-methylaniline	V
54	—	—	51	3,3′-dichloro-4,4′-dimethyl-azobenzene	V
55	cycloate	H	—	*N*-ethyl-*N*-cyclohexyl-*S*-ethyl-thiolcarbamate	IV
56	cycluron	H	—	3-cyclooctyl-1,1-dimethylurea	V
57	desmetryne	H	—	2-methylthio-4-methyl-amino-6-isopropylamino-*s*-triazine	XII
58	—	—	57	2-methylthio-4-methyl-amino-6-amino-*s*-triazine	XII
59	di-allate	H	—	*S*-2,3-dichloroallyl *N,N*-diisopropyl thiolcarbamate	IV
60	diazinon	I	—	*O,O*-diethyl *O*-(2-isopropyl-6-methyl-3-pyrimidyl) phosphorothioate	XI
61	—	—	60	*O,O*-diethyl *O*-(2-isopropyl-6-methyl-4-pyrimidyl) phosphate	XI
62	—	—	60	2-isopropyl-6-methyl-pyrimidine-4-ol	XI
63	dichlobenil	H	50	2,6-dichloro-benzonitrile	III
64	—	—	50, 63	2,6-dichloro-benzamide	III
65	dichlofluanid	F	—	*N*-(dichlorofluoromethyl-thio)-dimethylamino-sulfonic-anilide	VII
66	—	—	65	dimethylaminosulfonic anilide	VII
67	dicrotophos	I	—	*O,O*-dimethyl *O*-(1-methyl-3-dimethylamino-3-oxo-propene-1-yl) phosphate	III

Table XX. (*Continued*)

No.	Common name	Use[a]	Degradation product of	Chemical name	Retention time in Table
68	diethquinalphion	I	—	*O,O*-diethyl *O*-(quinoxaline-2-yl) phosphorothioate	XI
69	—	—	68	*O,O*-diethyl *O*-(quinoxaline-2-yl) phosphate	XI
70	dimefox	I	—	bis(dimethylamido) phosphoryl fluoride	VII
71	dimethoate	I	90	*O,O*-dimethyl *S*-(*N*-methylcarbamoylmethyl) phosphorodithioate	III
72	omethoate (= dimethoate-met)	I	71, 90	*O,O*-dimethyl *S*-(*N*-methylcarbamoylmethyl) phosphorothioate	III
73	dimetilan	I	—	2-dimethylcarbamoyl-3-methylpyrazolyl-(5)-dimethyl carbamate	XI
74	dinoseb	H	24, 75	2-*sec*-butyl-4,6-dinitrophenol	VIII
75	dinoseb acetate	H	—	2-*sec*-butyl-4,6-dinitrophenyl acetate	VIII
76	dioxacarb	I	—	2-(1,3-dioxalan-2-yl) phenyl *N*-methylcarbamate	IV
77	diuron	H	—	3-(3,4-dichlorophenyl)-1,1-dimethylurea	V
78	DNOC	H	—	2-methyl-4,6-dinitrophenol	VIII
79	dursban	I	—	*O,O*-diethyl *O*-(3,5,6-trichloro-2-pyridyl) phosphorothioate	X
80	fenazaflor	I	—	5,6-dichloro-1-phenoxycarbonyl-2-trifluoromethyl benzimidazole	XI
81	fenitrothion	I	—	*O,O*-dimethyl *O*-(3-methyl-4-nitrophenyl) phosphorothioate	VIII
82	fluometuron	H	—	3-(3-trifluoromethylphenyl)-1,1-dimethylurea	V
83	—	—	82	3-(3-trifluoromethylphenyl)-1-methylurea	V
84	—	—	82	3-(3-trifluoromethylphenyl)-urea	V
85	—	—	82	3-trifluoromethylaniline	V
86	—	—	82	3,3′-bis-trifluoromethyl-azobenzene	V
87	fluorodifen	H	—	2,4′-dinitro-4-trifluoro-methyl-diphenyl ether	VIII
88	—	—	87, 130, 132	4-nitrophenol	VIII
89	folpet	F	—	*N*-(trichloromethylthio) phthalimide	X

Table XX. (*Continued*)

No.	Common name	Use[a]	Degradation product of	Chemical name	Retention time in Table
90	formothion	I	—	*O,O*-dimethyl *S*-(*N*-formyl-*N*-methylcarbamoyl-methyl) phosphoro-dithioate	III
91	fuberidazole	F	—	2-(2′-furyl)-benzimidazole	XI
92	ioxynil	H	—	3,5-diiodo-4-hydroxy-benzonitrile	III
93	lenacil	H	—	3-cyclohexyl-5,6-trimethylene-uracil	XI
94	linuron	H	—	3-(3,4-dichlorophenyl)-1-methoxy-1-methylurea	V
95	—	—	77, 94, 126	3-(3,4-dichlorophenyl)-1-methylurea	V
96	—	—	77, 94, 126	3,4-dichloroaniline	V
97	—	—	77, 94, 126	3,3′,4,4′-tetrachloroazo-benzene	V
98	mercaptodimethur (methiocarb)	I	—	3,5-dimethyl-4-methylthio-phenyl *N*-methylcarbamate	IV
99	—	—	98	3,5-dimethyl-4-methyl-sulfonylphenyl *N*-methylcarbamate	IV
100	—	—	98	3,5-dimethyl-4-methyl-sulfonylphenyl *N*-methylcarbamate	IV
101	methabenzthiazuron	H	—	3-(2-benzothiazolyl)-1,3-dimethylurea	V
102	methidathion	I	—	*O,O*-dimethyl *S*[2-methoxy-1,3,4-thiadiazol-5(4H)-onyl-(4)-methyl] phosphorodithioate	XI
103	—	—	102	*O,O*-dimethyl *S*[2-methoxy-1,3,4-thiadiazol-5(4H)-onyl-(4)-methyl] phosphorothioate	XI
104	—	—	102	2-methoxy-1,3,4-thiadiazol-(4H)-5-on	XI
105	methoprotryne	H	—	2-methylthio-4-isopropyl-amino-6-(3-methoxypropyl-amino)-*s*-triazine	XII
106	—	—	105	2-methylthio-4-amino-6-(3-methoxypropylamino)-*s*-triazine	XII
107	metobromuron	H	—	3-(4-bromophenyl)-1-methoxy-1-methylurea	V
108	—	—	107	3-(4-bromophenyl)-1-methylurea	V
109	—	—	107	3-(4-bromophenyl) urea	V

Table XX. (*Continued*)

No.	Common name	Use[a]	Degradation product of	Chemical name	Retention time in Table
110	—	—	107	4-bromoaniline	V
111	—	—	107	4,4′-dibromoazobenzene	V
112	metoxuron	H	—	3-(3-chloro-4-methoxyphenyl)-1,1-dimethylurea	V
113	—	—	112	3-(3-chloro-4-methoxyphenyl)-1-methylurea	V
114	—	—	112	3-chloro-4-methoxyaniline	V
115	metribuzin	H	—	4-amino-6-*tert*-butyl-3-methylthio-1,2,4-triazine-5(4H)-on	XII
116	—	—	115	4-amino-6-*tert*-butyl-1,2,4-triazine-3,5(2H,4H)-di-on	XII
117	—	—	115	6-*tert*-butyl-3-methylthio-1,2,4-triazine-5(4H)-on	XII
118	—	—	115	6-*tert*-butyl-1,2,4-triazine-3,5(2H,4H)-di-on	XII
119	monalide	H	—	4-chloro-*N*-(α-keto-β,β-dimethyl)-*n*-amylaniline	III
120	monocrotophos	I	67	*O*,*O*-dimethyl *O*-(1-methyl-3-methylamino-3-oxo-propene-1-yl) phosphate	III
121	monolinuron	H	—	3-(4-chlorophenyl)-1-methoxy-1-methylurea	V
122	—	—	28, 121	3-(4-chlorophenyl)-1-methylurea	V
123	—	—	121	3-(4-chlorophenyl)-1-methoxyurea	V
124	—	—	28, 119, 121	4-chloroaniline	V
125	—	—	28, 119, 121	4,4′-dichloroazobenzene	V
126	neburon	H	—	3-(3,4-dichlorophenyl)-1-methyl-1-*n*-butylurea	V
127	nitralin	H	—	4-(methylsulfonyl)-2,6-dinitro-*N*,*N*-dipropylaniline	VIII
128	norflurazon	H	—	4-chloro-5-methylamino-2-(3′-trifluoromethylphenyl)-3(2H)-pyridazinon	XI
129	—	—	128	4-chloro-5-amino-2-(3′-trifluoromethylphenyl)-3(2H)-pyridazinon	XI
130	parathion	I	—	*O*,*O*-diethyl *O*-(4-nitrophenyl) phosphorothioate	VIII
131	—	—	130	*O*,*O*-diethyl *O*-(4-nitrophenyl) phosphate	VIII
132	parathion methyl	I	—	*O*,*O*-dimethyl *O*-(4-nitrophenyl) phosphorothioate	VIII
133	—	—	132	*O*,*O*-dimethyl *O*-(4-nitrophenyl) phosphate	VIII

Table XX. (*Continued*)

No.	Common name	Use[a]	Degradation product of	Chemical name	Retention time in Table
134	phenamiphos	N	—	*O*-ethyl-*O*-(3-methyl-4-methylthiophenyl) isopropylamino phosphate	VII
135	—	—	134	*O*-ethyl *O*-(3-methyl-4-methylsulfinylphenyl) isopropylamino phosphate	VII
136	—	—	134	*O*-ethyl *O*-(3-methyl-4-methylsulfonylphenyl) isopropylamino phosphate	VII
137	phenmedipham	H	—	(3-methoxycarbonylamino-phenyl)-*N*-(3′-methyl-phenyl) carbamate	IV
138	—	—	137	3,3′-dimethylazobenzene	IV
139	—	—	137	3,3′-dihydroxyazobenzene	IV
140	—	—	137	3-methyl-3′-hydroxyazo-benzene	IV
141	phosalone	I	—	*O,O*-diethyl *S*-(6-chloro-2-oxo-benzoxazolin-3-yl) methyl phosphorodithioate	X
142	phosphamidon	I	—	*O,O*-dimethyl *O*-(1-methyl-2-chloro-3-diethylamino-3-oxopropene-1-yl) phosphate	III
143	—	—	142	*O,O*-dimethyl *O*-(1-methyl-2-chloro-3-ethylamino-3-oxopropene-1-yl) phosphate	III
144	phoxim	I	—	*O,O*-diethyl *O*-(α-cyanobenzylidenamino) phosphorothioate	III
145	prometone	H	—	2-methoxy-4,6-bis-(iso-propylamino)-*s*-triazine	XII
146	prometryne	H	—	2-methylthio-4,6-bis-(iso-propylamino)-*s*-triazine	XII
147	propachlor	H	—	*N*-isopropyl α-chloro-acetanilide	III
148	propazine	H	—	2-chloro-4,6-bis-(iso-propylamino)-*s*-triazine	XII
149	propham	H	—	isopropyl *N*-phenylcarbamate	IV
150	propoxur	I	—	2-isopropoxyphenyl *N*-methylcarbamate	IV
151	pyrazon	H	—	5-amino-4-chloro-2-phenyl-3-pyridazone	XI
152	quintozene	F	—	pentachloronitrobenzene	VIII
153	—	—	152	pentachloroaniline	VII
154	simazine	H	—	2-chloro-4,6-bis(ethylamino)-*s*-triazine	XII

Table XX. (*Continued*)

No.	Common name	Use[a]	Degradation product of	Chemical name	Retention time in Table
155	terbuthylazine	H	—	2-chloro-4-ethylamino-6-*tert*-butylamino-*s*-triazine	XII
156	—	—	155	2-chloro-4-amino-6-*tert*-butylamino-*s*-triazine	XII
157	terbutryne	H	—	2-methylthio-4-ethylamino-6-*tert*-butylamino-*s*-triazine	XII
158	—	—	157	2-methylthio-4-amino-6-*tert*-butylamino-*s*-triazine	XII
159	thiophanate	F	—	1,2-bis-(3-ethoxycarbonyl-2-thioureido)benzene	VII
160	tri-allate	H	—	*S*-2,3,3-trichloroallyl *N*,*N*-diisopropyl thiolcarbamate	IV
161	tridemorph	F	—	2,6-dimethyl-4-tridecylmorpholine	X
Nitrogen-free organophosphorus pesticides:					
162	malathion	I	—	*O*,*O*-dimethyl *S*-[1,2-bis-(ethoxy-carbonyl)ethyl] phosphorodithioate	IX
163	mevinphos	I	—	*O*,*O*-dimethyl *O*-(2-methoxycarbonyl-1-methylvinyl) phosphate	IX
164	fenthion	I	—	*O*,*O*-dimethyl *O*-(3-methyl-4-methylthiophenyl) phosphorothioate	IX

[a] F = fungicide, H = herbicide, I = insecticide and/or acaricide, and N = nematicide.

packings. The comparison of the retention times of the aforementioned seven pairs of substances on the SE 30 column and on the QF 1 column demonstrates the examples. These examples show the appropriateness of the use of two-column or multicolumn instruments with different column packings in pesticide analysis.

However, the pairs diuron-neburon and atrazin(met)-chlorpropham in Table XIX show corresponding retention times as well on the SE 30 column as on the QF 1 column. In such cases it is, for a satisfactory identification, necessary to try other column temperatures, other column lengths, or other column packings, or even to use specific detectors or other analytical methods such as thin-layer chromatography or infrared or mass spectroscopy.

Our former references to the versatile usage of the *N*-FID in glc herbicide analysis (Maier-Bode 1971, Table 188) can be extended

Table XXI. *Trade names of pesticides mentioned in text.*

Trade name[a]	Number[b]	Tested degradation products (number)[c]
Accothion	81	—
Acricid	24	74
Actril	92	—
Afalon	94	95, 96, 97
Aflix	90	71, 72
Agermin	149	—
Alipur	39 + 56	—
Anilix	45	—
Anthio	90	71, 72
Antinonnin	78	—
Aresin	121	122, 123, 124, 125
Aretit	75	74
Avadex	59	—
Avadex BW	160	—
Azodrin	120	—
Barusil	68	69
Basudin	60	61, 62
Baycid	164	—
Baygon	150	—
Bayrusil	68	69
Baytex	164	—
Baythion	144	—
Bentrol	92	—
Betanal	137	138, 139, 140
Bidrin	67	120
Birgin	149	—
Bladan	130	88, 131
Blattanex	150	—
Botrysan	8	—
Brasoran	20	—
Brassicol	152	153
Calixin	161	—
Caparol	146	5, 6
Carbicron	67	120
Carbyne	22	—
Casoron	63	64
Cercobin	159	—
Certrol	92	—
CIPC	49	—
Cotoran	82	83, 84, 85, 86
Cottoran	82	83, 84, 85, 86
Cyfen	81	—
Cygon	71	72
Cythion	162	—
Dalf	132	88, 133
Dicuran	51	52, 53, 54
Dimecron	142	143
Dimetate	71	72
Dimethoat	71	72
DMU	77	95, 96, 97

Table XXI. (*Continued*)

Trade name[a]	Number[b]	Tested degradation products (number)[c]
DNBP	74	—
Dosanex	112	113, 114
Dursban	79	—
Dyrene	8	—
E 605	130	88, 131
Ekatox	130	88, 131
Ektafos	67	120
Elocron	76	—
Elvaron	65	66
Endosan	24	74
Entex	164	—
Eptapur	28	—
Euparen	65	66
Evik	3	4, 5, 6
Famid	76	—
Faneron	26	27
Fargo	160	—
Folidol	130	88, 131
Folidol M	132	99, 133
Folimat	72	—
Folithion	81	—
Folosan	152	153
Fostion MM	71	72
Fundal	40	41, 42, 43, 44
Galecron	40	41, 42, 43, 44
Gardoprim	155	14, 156
Gatnon	23	—
Gebutox	74	—
Gesagard	146	5, 6
Gesamil	148	15
Gesapax	3	4, 5, 6
Gesaprim	13	14, 15
Gesaran	105	5, 6, 106
Gesatamin	9	10, 11, 12
Gesatop	154	14
Gusathion	18	19
Gusathion A	16	17
Gusathion K	16	17
Gusathion M	18	19
Guthion	18	19
Hedolit	78	—
Hyvar	25	—
Igran	157	4, 6, 158
IPC	149	—
Ivosit	75	74
Kaptan	31	—
Karmex	77	95, 96, 97
Kloben	126	95, 96, 97
Lebaycid	164	—
Lorox	94	95, 96, 97

Table XXI. (*Continued*)

Trade name[a]	Number[b]	Tested degradation products (number)[c]
Lovozal	80	—
Lutin	78	—
Malathion	162	—
Malathon	162	—
Malathiozol	162	—
Maloran	35	36, 37, 38
Manzaflor	80	—
Matacil	7	—
ME 605	132	88, 133
Mercaptophos	164	—
Merpelan	23 + 93	—
Mesoranil	20	—
Mesurol	98	99, 100
Metacide	132	88, 133
Milbex	45	—
Milogard	148	15
Monitor	2	—
Morestan	34	—
Morocide	24	74
Nemacur	134	135, 136
Nemacur P	134	135, 136
Neocidol	60	61, 62
Niran	130	88, 131
Nuvacron	120	—
Orthene	1	2
Orthocid	31	—
Padan	33	—
Pantrin	32	—
Parathion	130	88, 131
Parathion-methyl	132	88, 131
Patoran	107	108, 109, 110, 111
PCNB	152	153
PD 5	163	—
Perfekthion	71	72
Pestox XIV	70	—
Phaltan	89	—
Phosdrin	163	—
Planavin	127	—
Potablan	119	124, 125
POX	130	88, 131
Pramitol	145	11, 12
Prebane	157	4, 6, 158
Prefix	50	63, 64
Preforan	87	88
Premerge	74	—
Primatol	145	11, 12
Primatol A	13	14, 15
Primatol S	154	14
Pyramin	151	—
Quinoxaline	34	—

Table XXI. (*Continued*)

Trade name[a]	Number[b]	Tested degradation products (number)[c]
Ramrod	147	—
Raphatox	78	—
Rogor	71	72
Ro-Neet	55	—
Roxion	71	72
Rubitox	141	—
Selinon	78	—
Semeron	57	5, 6, 58
Sencor	115	116, 117, 118
Sevin	32	—
Sinox	78	—
Snip	73	—
Sumithion	81	—
Supracid	102	103, 104
Sutan	30	—
Tamaron	2	—
Tenoran	46	47, 48
Terra-Systam	70	—
Terraclor	152	153
Thiophos	130	88, 131
Tixit	149	—
Topsin	159	—
Trevespan	92	—
Tribunil	101	—
Tritisan	152	153
Ultracid	102	103, 104
Unden	150	—
Valexon	144	—
Venzar	93	—
Voronit	91	—
Zolone	141	—
Zorial	128	129

[a] Trade names are quoted from the "Pesticide Manual" of the British Crop Protection Council (H. MARTIN 1972) and from the Pflanzenschutzmittel-Verzeichnis of the Biologische Bundesanstalt für Land- und Forstwirtschaft, Braunschweig (23rd edition 1973). Most of these trade names are patented as registered trade-marks and, therefore, their free use is prohibited. Not all trade names of the formulations of each active compound could be mentioned, this being no valuation whatsoever of the commercial product concerned. The national registers, for instance the Pflanzenschutzmittel-Verzeichnis of the Biologische Bundesanstalt für Land- und Forstwirtschaft, Braunschweig, give information on further trade names which are not mentioned in this table.

[b] The active compounds of the commercial products are marked in the text with the numbers mentioned in this rubric. They are identical with the numbers of rubric 1 in Table XX.

[c] This rubric indicates the degradation products of the active compounds in rubric (b). Their glc retention times are in Tables III to V and VII to XIX. Common names and chemical names of the active compounds as well as chemical names of the degradation products with the corresponding numbers are in Table XX.

to numerous pesticides containing nitrogen and also containing phosphorus, as this review shows. The main application areas are residue analysis, including multiresidue and forensic analysis, metabolism research, and formulation analysis as well as purity control.

Acknowledgments

To the Bundesministerium für Bildung und Wissenschaft, Bonn, to the Landesamt für Forschung des Landes Nordrhein-Westfalen, Düsseldorf, and to the Deutsche Forschungsgemeinschaft, Bonn-Bad Godesberg, the senior author expresses thanks for financial support for these evaluations, to the Badische Anilin- & Soda-Fabrik AG, Ludwigshafen; Bayer AG, Leverkusen; C. H. Boehringer Sohn, Ingelheim am Rhein; Ciba-Geigy AG, Basel; Consortium für Elektrochemische Industrie GmbH, München; Farbwerke Hoechst AG, Frankfurt (Main); Fisons Limited, Harston, Cambridge; Sandoz A.G., Basel; and to Schering AG, Berlin, for the submission of samples of the pesticides and metabolites studied.

Summary

More than half of the presently used pesticides contain organic nitrogen, therefore, it is possible that organic nitrogen compounds may occur in food and environment as residues of the pesticide application. Nearly all of these organic nitrogen compounds can be identified and quantitatively determined by gas chromatography (glc).

The nitrogen flame ionization detector (*N*-FID) has proved to be extremely appropriate, for cleanup is simplified by the selectivity of this detector and the sensitivity reaches a detection limit of picogram quantities. The construction of this detector, its operating parameters, and the detecting mechanism are described in detail.

This review contains particulars and tables covering the degradation, metabolism, toxicology, analysis, and the glc behaviour of 161 organic nitrogen-containing compounds which may possibly occur as components of pesticide residues, among them 94 fungicide, herbicide, insecticide, acaricide, and nematicide active compounds and 67 of their degradation products or metabolites. They include amides, thioamides, nitriles, carbamates, thiocarbamates, urea derivatives, amino compounds, nitrobenzene derivatives, and heterocyclic compounds with one, two, and three ring-nitrogen atoms in the molecule.

This review shows that two-column or multicolumn instruments with different column packings, eventually with detectors of differing selectivity, are necessary for satisfactory glc identification and evaluation.

References

ANLIKER, R., E. BERIGER, M. GEIGER, and K. SCHMID: Über die Synthese von Phosphamidon und seinen Abbau in Pflanzen. Helvet. Chimica Acta **44**, 1622 (1961).

ANONYMOUS: Changes in official methods of analysis. Captan. 84th Ann. Meeting, Oct. 12–15, 1970. J. Assoc. Official Anal. Chemists **54**, 451 (1971).

AUE, W. A., G. W. GEHRKE, R. C. TINDLE, D. L. STALLING, and D. D. RUYLE: Application of the alkali flame detector to nitrogen-containing compounds. J. Gas Chromatog. **5**, 381 (1967).

BACHE, C. A., and D. J. LISK: Determination of oxidative metabolites of dimethoate and thimet in soil by emission spectroscopic gas chromatography. J. Assoc. Official Anal. Chemists **49**, 647 (1966).

BAUNOK, I., and H. GEISSBÜHLER: Specific determination of urea herbicide residues by EC gas chromatography after hydrolysis and iodine derivate formation. Bull. Environ. Contam. Toxicol. **3**, 7 (1968).

BECKMAN, H., and D. GARBER: Recovery of 65 organophosphorus pesticides from Florisil with a new solvent elution system. J. Assoc. Official Anal. Chemists **52**, 286 (1969).

BEVENUE, A., and J. N. OGATA: Determination of bromacil by gas chromatography. J. Chromatog. **46**, 110 (1970).

BEYNON, K. J., and A. N. WRIGHT: The fates of the herbicides chlorthiamid and dichlobenil in relation to residues in crops, soils, and animals. Residue Reviews **43**, 23 (1972).

——, L. DAVIES, and K. ELGAR: Analysis of crops and soils for residues of 2,6-dichlorobenzonitrile (dichlobenil) and 2,6-dichlorothiobenzamide (chlorthiamid). II. Results. J. Sci. Food Agr. **17**, 156 (1966a).

—— —— ——, and A. N. WRIGHT: Analysis of crops and soils for residues of 2,6-dichlorobenzonitrile (dichlobenil) and 2,6-dichlorothiobenzamide (chlorthiamid). I. Development of method. J. Sci. Food Agr. **17**, 151 (1966b).

BOGGS, H. M.: Gas chromatography of dinitro herbicides. J. Assoc. Official Anal. Chemists **49**, 772 (1966).

BONELLI, E. J., H. HARTMANN, and K. P. DIMICK: Gas chromatography retention times and sensitivity data for insecticides and herbicides. J. Agr. Food Chem. **12**, 332 (1964).

BOWMAN, M. C., and M. BEROZA: Gas chromatographic analysis of 3-hydroxy-N-methyl-cis-crotonamide dimethyl phosphate (Azodrin) and 3-hydroxy-*N*,*N*-dimethyl-*cis*-crotonamide dimethyl phosphate (Bidrin). J. Agr. Food Chem. **15**, 465 (1967).

—— —— Determination of mesurol and five of its metabolites in apples, pears, and corn by gas chromatography. J. Assoc. Official Anal. Chemists **52**, 1054 (1969).

—— ——, and K. R. HILL: Chromatograms of foods for multicomponent residue determinaton of pesticides containing phosphorus and/or sulfur by GLC with flame photometric detection. J. Assoc. Official Anal. Chemists **54**, 346 (1971).

BRAZHNIKOV, V. V., M. V. GUR'EV, and K. I. SAKODYNSKY: Thermionic detectors in gaschromatography. Chromatog. Rev. **12**, 1 (1970).

BURKE, J. A., and L. GIUFFRIDA: Investigations of electron capture gas chromatography for the analysis of multiple chlorinated pesticide residues in vegetables. J. Assoc. Official Anal. Chemists **47**, 326 (1964).

——, and W. HOLSWADE: A gas chromatographic column for pesticide residue analysis: Retention times and response data. J. Assoc. Official Anal. Chemists **49**, 374 (1966).

Ciba-Geigy AG, Basle, Agricultural Division: Personal communication (1970).
CLEMONS, G. P., and R. E. MENZER: Oxidative metabolism of phosphamidon in rats and a goat. J. Agr. Food Chem. **16**, 312 (1968).
COFFIN, D. E.: Oxidative metabolism and persistence of parathion and malathion on field-sprayed lettuce. J. Assoc. Official Anal. Chemists **49**, 1018 (1966).
COULSON, D. M.: Electrolytic conductivity detector for gas chromatography. J. Gas Chromatog. **3**, 134 (1965).
—— Selective detection of nitrogen compounds in electrolytic conductivity gas chromatography. J. Gas Chromatog. **4**, 285 (1966 a).
—— Electrolytic conductivity detection for gas chromatography. Adv. Chromatog. **3**, 197 (1966 b).
CREMER, E.: New selective detectors in gas chromatography. J. Gas Chromatog. **5**, 329 (1967).
Deutsche Forschungsgemeinschaft: Rückstandsanalytik von Pflanzenschutzmitteln. Loseblattsammlung. Weinheim/Bergstasse: Verlag Chemie (1972).
DRÄGER, G.: Gaschromatographische Methode zur Bestimmung von ®Bayrusil-Rückständen in Pflanzenmaterial. Pflanzenschutz-Nachrichten BAYER **22**, 318 (1969).
EBERLE, D. O., and W. D. HÖRMANN: Fate of S-[(2-methoxy-5-oxo-Δ^2-1,3,4-thiadiazolin-4-yl)-methyl]-*O*,*O*-dimethyl phosphorodithioate (Supracide) in field-grown agricultural crops and soil. J. Assoc. Official Anal. Chemists **54**, 150 (1971).
——, and D. NOVAK: Fate of diazinon in field-sprayed agricultural crops, soil, and olive oil. J. Assoc. Official Anal. Chemists **52**, 1067 (1969).
EBING, W.: Gaschromatographischer Rückstandsnachweis von 47 phosphorhaltigen Insektizid-Wirkstoffen nach einem Einheitsverfahren. Pflanzenschutzber. **38**, 1 (1968).
—— Problem der Rückstandsanalytik von Bioziden aus der Sicht des Umweltschutzes. Deutsche Lebensmittel-Rundschau **69**, 39 (1973).
EL-REFAI, A., and T. L. HOPKINS: Parathion absorption, translocation, and conversion to paraoxon in bean plants. J. Agr. Food Chem. **14**, 588 (1966).
ESSER, H. O., G. DUPUIS, W. MÜCKE, and P. W. MÜLLER: Metabolism of the insecticide GS 13005. Presented VIth Internat. Congress Plant Prot., Vienna, Aug. 30–Sept. 6 (1967).
Farbwerke Hoechst AG, Analytic Laboratory: Personal communication (1972).
GEISSBÜHLER, H., and G. VOSS: Metabolism of substituted urea herbicides. Pesticide terminal residues. Presented 1971 IUPAC-Symposium, Tel-Aviv, p. 305. London: Butterworth (1971).
——, K. KOSSMANN, I. BAUNOK, and V. F. BOYD: Determination of total residues of chlorphenamidine [*N*′-(4-chloro-*o*-tolyl)-*N*,*N*-dimethylformamidine] in plant and soil material by colorimetry and thin-layer and electron capture gas chromatography. J. Agr. Food Chem. **19**, 365 (1971).
GIANG, B. Y., and H. F. BECKMAN: Determination of Bidrin, Azodrin, and their metabolites with the thermionic detector. J. Agr. Food Chem. **16**, 899 (1968).
GORBACH, S., and U. WAGNER: Pentachloronitrobenzene residues in potatoes. J. Agr. Food Chem. **15**, 654 (1967).
GRAETZ, D. A., G. CHESTERS, and T. C. DANIEL: Parathion degradation in lake sediments. J. Water Pollut. Control Fed. **42**, R76 (1970).
GUARDIGLI, A., W. CHOW, P. M. MARTWINSKI, and M. S. LEFAR: Determination of phosalone and its oxygen analog in citrus crops. J. Agr. Food Chem. **19**, 742 (1971).
GUTENMANN, W. H., and D. J. LISK: Gas chromatographic determination of phenolic pesticides and residues. J. Assoc. Official Anal. Chemists **48**, 1173 (1965).
HACSKAYLO, J., and D. L. BULL: Metabolism of dimethoate in cotton leaves. J. Agr. Food Chem. **11**, 464 (1963).

HARTMANN, H.: Phosphorus detector for pesticide analysis. Bull. Environ. Contam. Toxicol. **1**, 159 (1966).

HOLDEN, E. R., W. M. JONES, and M. BEROZA: Determination of residues of methyl- and dimethylcarbamate insecticides by gas chromatography of their 2,4-dinitroaniline derivatives. J. Agr. Food Chem. **17**, 56 (1969).

HOWARD, S. F., and G. YIP: Diazomethane methylation of a mixture of chlorophenoxy acids and dinitrophenols. J. Assoc. Official Anal. Chemists **54**, 970 (1971).

HRIVNÁK, J., and Z. STOTA: Gas chromatography of free phenolic pesticides. J. Gas Chromatog. **6**, 9 (1968).

IVES, N. F., and L. GIUFFRIDA: Investigation of thermionic detector response for the gas chromatography of P, N, As, and Cl organic compounds. J. Assoc. Official Anal. Chemists **50**, 1 (1967).

JANAK, J., V. SVOJANOVSKY, and M. DRESSLER: Analytical performance of an all-glass version of the coupled flame ionization-sodium thermionic detector. Czech. Chem. Commun. **33**, 740 (1968).

JOLLIFFE, V. A., B. E. DAY, L. S. JORDAN, and J. D. MANN: Method of determining bromacil in soils and plant tissues. J. Agr. Food Chem. **15**, 174 (1967).

JORDAN, L. S., W. J. FARMER, J. R. GOODIN, and B. E. DAY: Nonbiological detoxication of the *s*-triazine herbicides. Residue Reviews **32**, 267 (1970).

KARMEN, A.: Specific detection of halogen and phosphorus by flame ionization. Anal. Chem. **36**, 1416 (1964).

——, and L. GIUFFRIDA: Enhancement of the response of the hydrogen flame detector to compounds containing halogens and phosphorus. Nature **201**, 1204 (1964).

KATZ, S. E., and C. A. FASSBENDER: Chromatographic determination of metobromuron and metabolites in soil. Weed Sci. **16**, 401 (1968).

——, and R. F. STRUSZ: Chromatographic procedure for the determination of Maloran and metabolites in soils. Bull. Environ. Contam. Toxicol. **3**, 258 (1968).

—— —— Gas chromatographic separation of several urea herbicides and their metabolites. J. Agr. Food Chem. **17**, 1409 (1969).

KAUFMAN, D. D., and P. C. KEARNEY: Microbial degradation of *s*-triazine herbicides. Residue Reviews **32**, 235 (1970).

KNUESLI, E., D. BERRER, G. DUPUIS, and H. ESSER: *s*-Triazines. *In* P. C. KEARNEY and D. D. KAUFMAN (eds.): Degradation of herbicides. New York: Dekker, (1969).

KOSSMANN, K.: Methoden zur Rückstandsbestimmung von Phenmedipham in Pflanzenmaterial. Weed Res. **10**, 340 (1970).

——, H. GEISSBÜHLER, and V. F. BOYD: Specific determination of chlorphenamidine [*N'*-(4-chloro-*o*-tolyl)-*N*,*N*-dimethylformamidine] and of some potential metabolites in plant material by thin-layer and flame ionization gas chromatography. J. Agr. Food Chem. **19**, 360 (1971).

KRAUSE, Chr., and J. KIRCHHOFF: Gaschromatographische Bestimmung von Organophosphatrückständen auf Marktproben von Obst und Gemüse. Deutsche Lebensmittel-Rundschau **66**, 194 (1970).

KRISHNA, J. G., H. W. DOROUGH, and J. E. CASIDA: Synthesis of *N*-methylcarbamates via methyl isocyanate-C^{14} and chromatographic purification. J. Agr. Food Chem. **10**, 462 (1962).

KUCHAR, E. J., F. O. GEENTY, W. P. GRIFFITH, and R. J. THOMAS: Analytical studies of metabolism of Terraclor in beagle dogs, rats, and plants. J. Agr. Food Chem. **17**, 1237 (1969).

LASKI, R. R., and R. R. WATTS: Gas chromatography of organonitrogen pesticides, using a nitrogen-specific detection system. J. Assoc. Official Anal. Chemists **56**, 328 (1973).

LEMPERLE, E., and H. STRECKER: Gas-chromatographische Bestimmung des Wirkstoffrückstandes auf Weintrauben, in Traubenmost und in Wein nach Anwendung von Mycodifol. Z. anal. Chem. **253**, 275 (1971).

LUCIER, G. W., and R. E. MENZER: Metabolism of dimethoate in bean plants in relation to its mode of application. J. Agr. Food Chem. **16**, 936 (1968).

—— —— Nature of oxidative metabolites of dimethoate formed in rats, liver microsomes, and bean plants. J. Agr. Food Chem. **18**, 698 (1970).

MAIER-BODE, H.: Herbizide und ihre Rückstände. Stuttgart: Verlag Ulmer (1971).

MARTIN, H.: Pesticide manual. Basic information on the chemicals used as active components of pesticides, 3rd ed. Brit. Crop. Prot. Council (1972).

MARTIN, R. L.: Fast and sensitive method for determination of nitrogen. Anal. Chem. **38**, 1209 (1966).

MATTSON, A. M., R. A. KAHRS, and R. T. MURPHY: Routine quantitative residue determinations of *s*-[(2-methoxy-5-oxo-Δ^2-1,3,4-thiadiazolin-4-yl)methyl] *O,O*-dimethyl phosphorodithioate (Supracide) and its oxygen analog in forage crops. J. Agr. Food Chem. **17**, 565 (1969).

—— —— —— Quantitative determination of triazine herbicides in soils by chemical analysis. Residue Reviews **32**, 371 (1970).

MCKELLAR, R. L.: 2-*sec*-butyl-4,6-dinitrophenol and 2-amino-6-*sec*-butyl-4-nitrophenol in milk and cream from cows fed 2-*sec*-butyl-4,6-dinitrophenol. J. Agr. Food Chem. **19**, 758 (1971).

MCKONE, C. E.: The determination of some substituted urea herbicide residues in soil by electron-capture gas chromatography. J. Chromatog. **40**, 60 (1969).

——, and R. J. HANCE: Estimation of S-2,3,3-trichloroallyl *N,N*-diisopropylthiolcarbamate (triallate) residues in soil, barley straw, and grain by electron-capture gas chromatography. J. Agr. Food Chem. **15**, 935 (1967).

—— ——The gas chromatography of some substituted urea herbicides. J. Chromatog. **36**, 234 (1968).

MENDOZA, C. E., and J. B. SHIELDS: GLC analysis of Dyrene residues, using Chromosorb W (HP) with OV-17/QF-1 silicone coating. J. Assoc. Official Anal. Chemists **54**, 986 (1971).

MENGES, R. M., and J. L. HUBHARD: Selectivity, movement, and persistence of soil-incorporated herbicides in carrot plantings. Weed Sci. **18**, 247 (1970).

MENZER, R. E., and J. E. CASIDA: Nature of toxic metabolites formed in mammals, insects, and plants from 3-(dimethoxyphosphinyloxy)-*N,N*-dimethyl-cis-crotonamide and its *N*-methyl analog. J. Agr. Food Chem. **13**, 102 (1965).

METHRATTA, T. P., R. W. MONTAGNA, and W. P. GRIFFITH: Determination of Terraclor in crops and soil by electron-capture gas chromatography. J. Agr. Food Chem. **15**, 648 (1967).

MEULEMANS, K. J., and E. T. UPTON: Determination of dichlobenil and its metabolite, 2,6-dichloro-benzoic acid, in agricultural crops, fish, soil, and water. J. Assoc. Official Anal. Chemists **49**, 976 (1966).

MILES, J. R. W.: A new colorimetric method for determination of residues of Guthion and Ethyl Guthion and their oxygen analogs. J. Assoc. Official Anal. Chemists **47**, 882 (1964).

NEAL, R. A.: Metabolism of diethyl 4-nitrophenyl phosphorothioate (parathion) *in vitro*. Biochem. J. **103**, 183 (1967).

O'BRIEN, R. D.: Toxic phosphorus esters. New York: Academic Pre°s (1960).

OHLING, R. W., E. THALL, and P. H. OEY: General consideration concerning atmospheric aerosol monitoring with the hydrogen flame ionization detector. Anal. Chem. **41**, 302 (1969).

PEASE, H. L.: Determination of residues of 3-cyclohexyl-5,6-trimethyleneuracil. J. Sci. Food Agr. **17**, 121 (1966).

PECK, J. M., and K. J. HARKISS: Gas chromatographic analysis of some carbamate derivatives. J. Chromatogr. Sci. **9**, 370 (1971).

POMERANTZ, I. H., and R. ROSS: Captan and structurally related compounds by thin layer and gas-liquid chromatography. J. Assoc. Official Anal. Chemists **51**, 1058 (1968).

RENVALL, S., and M. ÅKERBLOM: Determination of organophosphorus pesticide residues in fruits and vegetables on the Swedish market from 1964 to 1968. Residue Reviews **34**, 1 (1971).

RIEDMANN, M.: Theoretical and practical aspects of thermionic detection of heteroelements. Presented 4th Internat. Symp. "Advances and Utilization of Chromatography," Bratislava (1973).

——, and C. B. EUSTON: New GC detector for phosphorus containing pesticides. Anal. Adv. *Hewlett-Packard Co.* Summer (1969).

RUZICKA, J., J. THOMSON, and B. B. WHEALS: The gas-chromatographic examination of organophosphorus pesticides and their oxidation products. J. Chromatog. **30**, 92 (1967).

Schering AG, Berlin, PFLANZENSCHUTZFORSCHUNG: Personal communication (1970).

SCHMIDT, K.-J., and I. HAMMANN: ®Bayrusil, ein neuer insektizider und akarizider Phosphorsäureester. Pflanzenschutz-Nachrichten BAYER **22**, 324 (1969).

SMITH, A. E.: Estimation of S-2,3-dichlorallyl *N,N*-diisopropylthiolcarbamate (diallate) residues in soils by electron-capture gas chromatography. J. Agr. Food Chem. **17**, 1052 (1969).

SPENGLER, D., and B. HAMROLL: Trennung und Bestimmung von Carbamat- und Harnstoff-Herbiziden durch Reaktions-Gaschromatographie. J. Chromatog. **49**, 205 (1970).

STORHERR, R. W., P. OTT, and R. R. WATTS: A general method for organophosphorus pesticide residues in nonfatty foods. J. Assoc. Official Anal. Chemists **54**, 513 (1971).

THORNTON, J. S., and C. A. ANDERSON: Determination of residues of Di-Syston and metabolites by thermionic emission flame gas chromatography. J. Agr. Food Chem. **16**, 895 (1968).

TINDLE, R. C., C. W. GEHRKE, and A. W. AUE: Improved GLC method for *s*-triazine residue determination. J. Assoc. Official Anal. Chem. **51**, 682 (1968).

VAN MIDDELEM, C. H., and R. E. WAITES: Gas chromatographic and colorimetric measurement of dimethoate residues. J. Agr. Food Chem. **12**, 178 (1964).

VERLOOP, A.: Fate of the herbicide dichlobenil in plants and soil in relation to its biological activity. Residue Reviews **43**, 55 (1972).

VOGELER, K., and H. GOELDNER: Untersuchungen über Rückstände nach Anwendung von Euparen an Weintrauben. Schweiz. Z. Obst- und Weinbau **103**, 494 (1967).

——, and H. NIESSEN: Gaschromatographische Bestimmung von Rückständen in Pflanzen nach Anwendung von ®Morestan. Pflanzenschutz-Nachrichten BAYER **20**, 530 (1967).

VOSS, G., and H. GEISSBÜHLER: Methodological aspects of urea herbicides residue analysis. Proc. 2nd Internat. IUPAC-Congress Pesticide Chemistry, Tel-Aviv, **4**, 525 (1971).

——, I. BAUNOCK, and H. GEISSBÜHLER: Phosphamidon residue methods. Residue Reviews **37**, 101 (1971).

WAGGONER, T. B.: Metabolism of Nemacur [ethyl 4-(methylthio)-*m*-tolyl isopropylphosphoramidate] and identification of two metabolites in plants. J. Agr. Food Chem. **20**, 157 (1972).

WALTER, J. P., M. L. KERCHERSID, and M. G. MERKLE: The gas chromatographic determination of 4-trifluoromethyl-2,4′-dinitrodiphenyl ether residues in soybeans. J. Agr. Food Chem. **16**, 143 (1968).

WATTS, R. R., and R. W. STORHERR: Gas chromatography of organophosphorus pesticides: Retention times and response data on three columns. J. Assoc. Official Anal. Chemists **52**, 513 (1969).

WESSEL, J. R.: Collaborative study of a method for multiple organophosphorus pesticide residues in nonfatty foods. J. Assoc. Official Anal. Chemists **50,** 430 (1967).

WESTLAKE, W. E., and F. A. GUNTHER: Advances in gas chromatographic detectors illustrated from applications to pesticide residue evaluations. Residue Reviews **18,** 175 (1967).

WHEELER, L., and A. STROTHER: Chromatography of *N*-methylcarbamates in the gaseous phase. J. Chromatog. **45,** 362 (1969).

WILLIAMS, R. T.: Detoxication mechanisms. The metabolism and detoxication of drugs, toxic substances and other organic compounds, 2nd ed. London: Chapman & Hall (1959).

WINDHAM, E. S.: Gas chromatographic column for pesticide analysis. J. Assoc. Official Anal. Chemists **52,** 1237 (1969).

WRIGHT, F. C., and J. C. RINER: Note on versatility of silicone Dow Corning 11 substrate for gas chromatography of pesticides. J. Assoc. Official Anal. Chemists **52,** 879 (1969).

YIP, G., and S. F. HOWARD: Extraction and cleanup procedure for the gas chromatographic determination of four dinitrophenolic pesticides. J. Assoc. Official Anal. Chemists **51,** 24 (1968).

ZIELINSKI, W. L., and L. FISHBEIN: Gas chromatography of carbamate derivatives. II. *N*-Substituted carbamates. J. Gas Chromatog. **3,** 333 (1965).

Manuscript received February 13, 1974; accepted April 15, 1974.

Subject Index